Taniguchi Symposia on Brain Sciences No.18

EMOTION, MEMORY AND BEHAVIOR

Studies on Human and Nonhuman Primates

Taniguchi Symposia on Brain Sciences No. 18
PROGRAM COMMITTEE

Teruo Nakajima (Chairman), P.S. Goldman-Rakic,
Taketoshi Ono, Tsukasa Koyama

The Taniguchi Foundation, Division of Brain Sciences
ORGANIZING COMMITTEE

Taniguchi Symposia on Brain Sciences No.18

EMOTION, MEMORY AND BEHAVIOR

Studies on Human and Nonhuman Primates

Edited by
Teruo Nakajima
and Taketoshi Ono

JAPAN SCIENTIFIC SOCIETIES PRESS Tokyo

CRC PRESS Boca Raton New York London Tokyo

Supported in part by the Ministry of Education, Science and Culture under Grant-in-Aid for Publication of Scientific Research Result.

Published by:
JAPAN SCIENTIFIC SOCIETIES PRESS
2-10 Hongo, 6-chome, Bunkyo-ku, Tokyo 113, Japan
ISBN 4-7622-8797-0

and

CRC PRESS
2000 Corporate Blvd., N.W., Boca Raton, FL 33431, U.S.A.
ISBN 0-8493-7777-3

Library of Congress Cataloging-in-Publication Data

Emotion, memory, and behavior: studies on human and nonhuman primates/edited by Teruo Nakajima, Taketoshi Ono.
p. cm.—(Taniguchi symposia on brain sciences; no. 18)
Includes bibliographical references and index.
ISBN 0-8493-7777-3 (alk. paper)
1. Neuropsychology—Congresses. 2. Emotions—Congresses. 3. Memory—Congresses. I. Nakajima, Teruo. II. Ono, Taketoshi. III. Series.
QP360. E466 1995 95-31217
612.8'2—dc20 CIP

Distributed in all areas outside Japan and Asia between Pakistan and Korea by CRC Press.

Printed in Japan

PREFACE

This edition is based on the 18th International Taniguchi Symposium on Brain Sciences held on the island of Hawaii, U.S.A., from February 5 to 8, 1995. When the Division of Brain Sciences of the Taniguchi Foundation was founded in 1976, Professor Osamu Hayaishi, the Chairman of the Organizing Committee, and I met the late Mr. Toyosaburo Taniguchi, the Foundation's sponsor, and the late Professor Kenjiro Shoda, an uncle of the Empress Michiko and the former director general of the foundation. We told them at that time that the purpose of the symposia of the division would be to nurture and assist young scientists in the development of basic brain science. As a neuropsychiatrist it was my view that neuropsychiatry had to become a more scientific discipline based on brain and behavioral sciences, so the advancement of these studies was critical for the neuropsychiatric clinics of the future. From this perspective, the theme of the 18th symposium was "Emotion, Memory and Behavior—Studies on Human and Nonhuman Primates", and the Program Committee selected promising young scientists in these two rapidly progressing fields from all over the world. I therefore believe that this book should be useful in furthering not only basic neuroscience but also neuropsychiatry in the years to come, and hope very much that it will initiate neurobiological studies for the appearance of mental symptoms such as anxiety, depression and so on.

The book is comprised of three parts. *Emotion* concerns the role of the limbic system in the demonstration of emotion and abnormal behavior, and details studies on an animal model of autism, the neuronal mechanism of emotion and behavior, and a PET study on depression. A neuropharmacological study on fear and anxiety (an animal model of anxiety) is also reviewed. *Memory* concerns the brain mechanisms of memory, and covers studies on the hippocampal place code, the neural mechanism of long-term memory in the temporal visual cortex, spatial and non-spatial memory (short-term memory) in the prefrontal cortex, and neuropsychological studies on amnesic patients. *Behavior* is concerned with the brain mechanisms of normal and abnormal behavior, and reports studies on understanding the behavior and 'mind' of others (visual processing within the temporal cortex), perception of geometric illusions, inhibition and facilitation of visual-motor links, self-mutilation behavior and a neuroanatomical study on cognitive aging. A neurochemical study on sleep and wakefulness is also of interest.

I am grateful to the Taniguchi Foundation, the Organizing Committee of the Division of Brain Sciences, the Program Committee of the symposium, and all participants for their valuable contributions. The editors believe this book will be found to be a current review of the status of these fields which should stimulate future studies.

June 1995 T. Nakajima

CONTENTS

Behavior

EMOTION

1

ONTOGENETIC DEVELOPMENT AND PLASTICITY OF COGNITIVE AND SOCIOEMOTIONAL PROCESSES

MAREE J. WEBSTER AND JOCELYNE BACHEVALIER*[1]

Laboratory of Neuropsychology, National Institute of Mental Health, Bethesda, MD 20892, U.S.A.

Research that has examined the cognitive and socioemotional development of monkeys with selective limbic*[2] and cortical lesions in infancy has revealed that, depending on the circumstances, infants can recover completely from early damage to neural systems, whereas in other circumstances there may be no improvement at all, and in still others the impairment may not become apparent until much later in development. This chapter summarizes several examples of these lesion effects and includes studies on brain plasticity that are yielding new insights into the neural mechanisms that underlie these various behavioral effects after early brain injury.

*[1]Present address: Department of Neurobiology and Anatomy, University of Texas Medical School, Houston, TX 77225, U.S.A.

*[2]In this report the term limbic refers to the medial temporal lobe structures that include the hippocampal formation, the amygdaloid complex, and adjacent entorhinal, perirhinal and parahippocampal cortex.

I. DEVELOPMENT OF MEMORY FUNCTIONS

Studies by Mishkin and colleagues, carried out on adult monkeys, indicated that stimulus memory and habit formation are two qualitatively different retention processes based on separate neural mechanisms (*25, 26, 30*). Subsequent studies carried out on normal infant monkeys demonstrated that these two systems are also dissociable developmentally. Whereas habit formation (as measured by concurrent object discrimination learning with 24-hour intertrial intervals, or 24-hour ITI task) is present in 3-month-old monkeys, cognitive memory (as measured by the delayed nonmatching-to-sample task, or DNMS) is immature at this stage and develops only slowly, not reaching adult levels until the animals are about 2 years of age (*1*). These findings suggested that the presumed substrate of habit formation, the cortico-striatal system, is operational in early infancy, whereas the substrate for stimulus memory, the cortico-limbic system, develops later in infancy, presumably because the circuitry on which it depends undergoes a relatively slow ontogenetic development.

Visual recognition memory (*i.e.* cognitive memory) is critically dependent on a neural circuit that includes both the inferior temporal cortex, the limbic structures, the diencephalic, and prefrontal areas to which the limbic structures project, as well as the basal forebrain and its feedback connections (for review see ref. *24*). To examine the possibility that immaturity of one or more of the neural stations in this pathway is responsible for the late development of this function, a series of experiments was conducted on the effects of early damage to either the limbic structures (including the hippocampal formation, amygdaloid complex, and adjacent rhinal cortical areas), known to be critical for the formation of new memories, or inferior temporal cortical area TE, which may be critical for the long-term storage of the visual memories. When tested at 10 months of age in the visual recognition task, infant monkeys with neonatal bilateral area TE lesions performed almost as well as normal infants, whereas the monkeys with neonatal limbic lesions were severely impaired (*5*) (Fig. 1). When compared to the severe cognitive memory loss found in adult monkeys following either area TE or limbic lesions, the findings

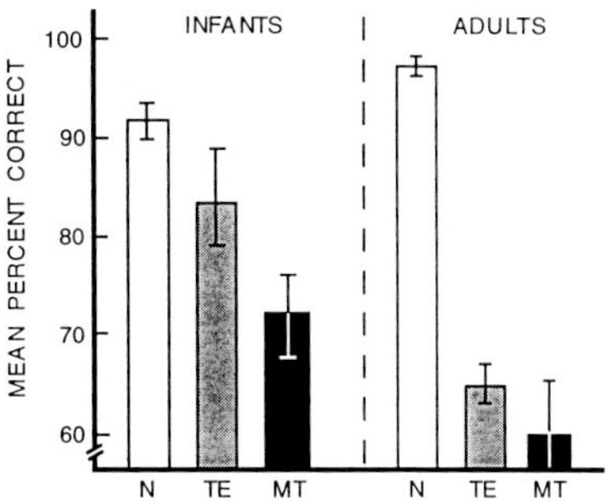

Fig. 1. Mean scores on the recognition performance test (averaged across delays and list lengths) of 10-month-old monkeys (operated neonatally) and adult monkeys. N, normal controls; TE, animals with bilateral TE lesions; MT, animals with bilateral medial temporal lobe lesions. (Adapted from ref. *1*)

in infants pointed to a greater compensatory potential after neonatal cortical than after neonatal limbic removal. These behavioral data implied both that the remaining, intact visual association areas outside of area TE possess considerable plasticity at birth and that, at the time of the early ablation, area TE also was still functionally immature. Direct evidence of neocortical immaturity in the macaque was provided by studies on the development of opiatergic and cholinergic receptors as well as on the maturation of metabolic activity (*7*, *8*, *27*), and also on the development of cortico-cortical and cortico-subcortical connections in the early postnatal period (*31*, *33*).

In the context of the proposed neural memory circuit, the data from developmental studies led to the conclusion that the poor memory ability shown by normal infant monkeys in DNMS was likely due to immaturity of inferior temporal cortical areas rather than immaturity of the limbic structures. Indeed, all the behavioral data gathered until now seems to indicate that early damage to limbic structures yields a behavioral pattern strikingly similar to that found after late damage. Thus, like limbic lesions in adulthood, limbic lesions in infancy result in a global and long-lasting anterograde amnesia, affecting not only memory for visual information but also memory for tactile stimuli and spatial locations (*20*). This memory loss was also apparent very early in life as revealed by the absence, in the infant monkeys with limbic lesions, of the normal infant monkey preference for viewing novel objects, and this deficit was still apparent when the animals reached adulthood (*6*). Apparently, limbic structures are operational very early in life enabling the early development of at least simple cognitive memory functions. The limiting factor in the development of more complex memory functions appears now to be the immaturity of higher order cortical areas, such as area TE.

II. BEHAVIORAL CONSEQUENCES OF EARLY DAMAGE TO MEDIAL TEMPORAL LOBE STRUCTURES *VS*. INFERIOR TEMPORAL CORTICAL AREA TE

To investigate whether the severe memory deficit following neonatal limbic lesions was accompanied by disturbances in emotional and social behavior, the social interactions between operated monkeys and their controls were analyzed at different points in their maturation. Thus, at the ages of 2 months, 6 months, 18 months, and 5 years, operated infant monkeys and their age-matched controls were placed in a play cage containing towels and toys, and the behavior of each pair was videorecorded for two periods of 5 min each, separated by a five minute interval, for 6 consecutive days. The frequency and duration of behaviors were scored independently by two observers who assigned the behaviors to one of 14 different behavioral categories, which included among others: approach, acceptance of approach, dominant approach, active withdrawal, inactivity, locomotor stereotypies, self-directed activities, and manipulation. (Interobserver reliability equalled or exceeded 0.90.) Compared with normal infant monkeys and those with neonatal cortical lesions, the monkeys with neonatal limbic lesions, at 2 months of age, had more temper tantrums when first placed in a novel play cage, showed more passive behavior, and manipulated objects less than the controls. Interestingly, at 6 months, these monkeys displayed even more striking pathology in that they now showed a complete lack of social contact, extreme submissiveness including active withdrawal, and gross motor stereotypies (Fig. 2). In addition, at 6 months, monkeys with limbic lesions displayed blank and unexpressive faces, few eye contacts, and poor body expression. These abnormalities were still present at 6–7 years of age, indicating that early damage to

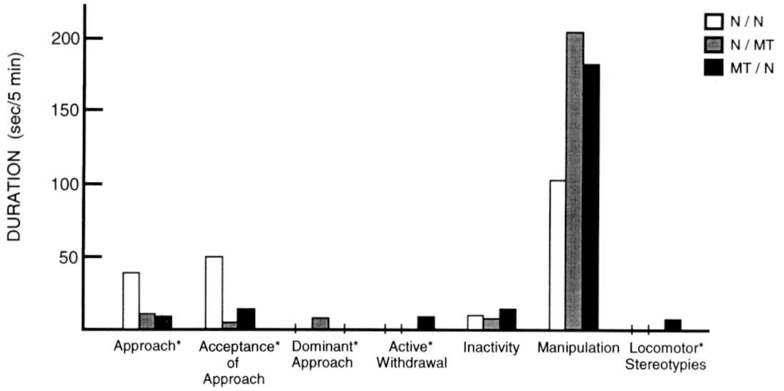

Fig. 2. Average duration (seconds per 5-minute session) of behavioral activities at 6 months of age in animals with neonatal medial temporal lobe damage (MT/N) and their age matched controls (N/MT) as compared with average duration of behavioral activities of normal controls paired together (N/N). Asterisk indicates statistical differences between groups ($p < .05$). Adapted from ref. *4*)

the limbic system yields long-lasting socioemotional disturbances (*19*). The developmental time course and the nature of these disturbances resemble those seen in autistic children (*2–4*). These neurobehavioral developmental studies, together with the recent reports documenting neuropathology in the amygdala and hippocampus of autistic subjects, provide support for the notion that early dysfunction of the limbic system may be one cause of infantile autism (*10*).

Although the animals that had received early damage to area TE showed none of the same socioemotional disturbances seen in infants with early limbic lesions, they did display other behavioral abnormalities (*e.g.* hyperactivity, and increased frequency of shifting from one behavioral activity to another) that resemble those of children with an attention deficit/hyperactivity disorder (*21*). These behavioral abnormalities were much less apparent when the animals with early TE damage reached adulthood (6–7 years) (*17, 18*).

III. BEHAVIORAL CONSEQUENCES OF NEONATAL DAMAGE TO THE AMYGDALA *VS.* THE HIPPOCAMPUS

In sum, the initial study had demonstrated that neonatal damage to the limbic structures in primates results in a profound loss in memory functions as well as severe socioemotional disorders. But is this to be considered a single complex syndrome or might it be fractionated by damaging specific components of the limbic system? Because there was considerable evidence from studies in adult monkeys that amygdala lesions, in particular, result in socioemotional disturbances, it was possible that the pathology responsible for the socioemotional developmental disturbances was not the combined amygdalohippocampal damage but rather the damage to the amygdaloid complex alone (*13, 15*). Conversely, because there was evidence from some investigators that hippocampal lesions, in particular, were the cause of memory loss (*16, 29*), the developmental amnesic syndrome following neonatal amygdalohippocampal damage might actually have resulted from damage to the hippocampal formation alone.

To investigate these possibilities newborn monkeys were prepared with damage to either the amygdaloid complex (including the amygdaloid nuclei and rostral portion of the entorhinal cortex) or the hippocampal formation (including the hippocampus and parahippocampal gyrus) and were tested in cognitive tasks and social interactions in the same way as were those monkeys with neonatal damage to the entire amygdalohippocampal complex.

Surprisingly, the results demonstrate that monkeys with neonatal hippocampal lesions are not impaired in recognition memory tasks (at least for delays up to 3 min) when tested at 10 months of age, whereas those with neonatal lesions of the amygdaloid complex are significantly impaired, though their impairment is not as severe as that of monkeys with the combined amygdalohippocampal lesions. Thus, either the amygdala and/or the adjacent rostral entorhinal cortex contribute more than the hippocampal formation to the development of cognitive memory (*3, 4*).

Early damage to the amygdaloid complex also resulted in a pattern of socioemotional disturbances similar to that found after the amygdalohippocampal lesions, although again the magnitude of these disturbances was much less (*3, 4*). By contrast, early damage to the hippocampal formation alone yielded only few effects, at least initially. At 2 months of age, these animals were less likely than the normal controls to initiate social contacts, they were somewhat less active than the controls, and they tended to display blank facial expressions. However, these few behavioral disturbances were not apparent when the animals reached 6 months of age, suggesting that they had recovered from their mild impairment (*3*, *4*). It was a surprise, therefore, to see the abnormal behaviors reemerge when these animals reached adulthood. The behavioral changes included a reduction again in social contacts, without changes in social status, and the appearance of locomotor stereotypies. The developmental time course and nature of these disturbances are somewhat reminiscent of those seen in schizophrenia (*11*).

It thus appears that the nature, degree, and time of appearance of the behavioral disturbances all depend critically upon the specific locus and extent of the neonatal limbic lesions. The exact substrates are still not clear, however, given the recent discovery that, in adults, rhinal cortical areas contribute substantially both to memory ability and to emotional responsivity (*22*, *23*). That finding raises the question of whether or not neonatal damage to rhinal cortical areas contributed to the effects we have seen. Studies of the behavioral consequences of neonatal damage restricted to rhinal cortical areas are clearly needed.

IV. PLASTICITY OF VISUAL MEMORY CIRCUITS

As described earlier in this chapter, a critical component of the neural circuitry that underlies visual recognition memory is the interaction of the occipitotemporal pathway and the limbic structures. Disrupting this interaction, by removing area TE, which is the final neural station in the occipitotemporal pathway and the major source of object-vision input to the limbic system, results in a severe impairment of visual recognition memory. However, as illustrated in Fig. 1, this is only true if the lesion is made in adulthood. If the lesion is made in infancy, monkeys demonstrate almost normal visual recognition memory. This sparing of visual memory function raises the question as to how does the visual cortical pathway interacts with the limbic structures after neonatal removal of area TE.

As a first step toward answering this question, we compared the anatomical organization of the visual memory pathways in normal infant and adult monkeys. We discovered that infant monkeys have far more widespread corticocortical projections emanating from inferior temporal areas TE and TEO than do adults (*31*) (Fig. 3). For example, infant monkeys have projections from area TE to perirhinal area 35 as well as to parahippocampal areas TF and TH, and also from area TEO to area TF, none of these being apparent in adults. Moreover, while area TE projects to area 36 in both adult and infant monkeys, the projection was found to be much more widespread in the infants, both in areal extent and in laminar distribution. We also found more extensive projections from area TE to various subcortical structures in infant monkeys, including the superior colliculus and nucleus medialis dorsalis of the thalamus, as well as a projection from

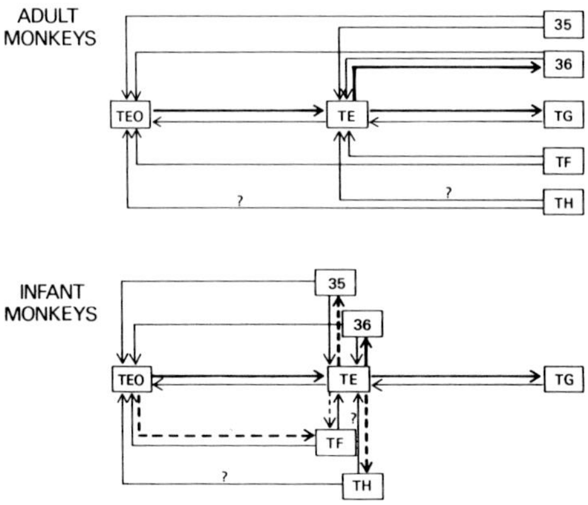

Fig. 3. Summary diagram illustrating the feedforward and feedback projections in adult and infant monkeys based on laminar patterns of connections following injections of retrograde and anterograde tracers in areas TEO and TE. Heavy solid lines indicate feedforward projections, *i.e.*, those originating in layer III and/or terminating mainly in layer IV. Thin solid lines indicate feedback projections, *i.e.*, those projections originating mainly in layers V/VI and/or terminating outside layer IV. Broken lines denote transient projections that are present in infants but not in adults. Question marks indicate those projections that originate from both layers III and V/VI and thus may turn out to be of the intermediate type rather than feedback. As one progresses from left to right in the figure, there is a progression from lower-order to higher-order areas; areas within a column are at the same hierarchical level. Thus, in the adult, areas 35, 36, TG, TF, and TH are higher-order than area TE, whereas in the infant, areas 35, 36, TF, and TH are at the same hierarchical level as area TE. The adult pattern is achieved by the removal of transient projections from area TEO to area TF, from area TE to areas 35, TF and TH, and from area TE to layer I of area 36. (Reproduced from ref. *31*)

area TEO to the lateral basal nucleus of the amygdala (*33*). Because all of these projections are either absent or more restricted in adults than in infants, they must retract during normal development.

The presence of transient projections from area TEO to both the amygdala and parahippocampal cortex suggested a possible anatomical pathway, present in infants but not in adults, to account for the sparing of memory after removal of area TE in infancy. To test the hypothesis that these transient projections are retained after neonatal removal of area TE, and could thereby possibly mediate the sparing of visual memory, we examined connections of area TEO to the amygdala and parahippocampal cortex in monkeys that had received area TE lesions in infancy or in adulthood. We found that, in fact, the transient TEO-amygdalar projection does survive into adulthood if area TE is removed neonatally (*32*). Thus, the area TEO projection, which normally retracts during development, becomes stabilized at terminal sites in the amygdala that have been vacated by the removal of area TE projections. Moreover, removal of area TE neonatally induces the expansion (*i.e.* sprouting) of a normally limited projection from TEO to the dorsal part of the lateral nucleus of the amygdala, where it invades the terminal space normally occupied by terminals from area TE. While some of the normally transient corticocortical connections of area TEO are maintained following the neonatal TE lesion, not all are. Among those lost, just as in the normal infant, are projections from area TEO to TF. Also, after the neonatal TE lesion, there is no sprouting of projections from TEO into the perirhinal area, like there is into the amygdala. At the same time, in contrast to animals that had area TE removed in infancy, we found absolutely no reorganization of connections in monkeys that had area TE removed in adulthood (*32*).

To determine if the maintenance and sprouting of projections from area TEO to the amygdala could have played a role in the permanent preservation of visual memory ability in monkeys that had received bilateral removal of area TE in infancy, we subsequently removed area TEO in four such monkeys. First, the monkeys were tested on the recognition memory task, DNMS, at 2 years of age, prior to the secondary lesion, to confirm the long-term preservation of spared memory function, and then they were given the secondary

TEO lesion and retested. Surprisingly, their performance did not decline. The negative finding suggests either that the maintained TEO-limbic connections are not participating in visual recognition but serve other, perhaps visuo-emotional, functions, or that they do participate in recognition but that additional areas are also contributing to this memory function.

Indeed, in adult monkeys, cortical areas other than area TE are known to receive visual information and to project to limbic structures, though these areas are not normally important for visual recognition memory. However, they may be important for this function in infant monkeys, or they may become important if area TE is removed in infancy. The most likely candidates for such a role are: 1) the superior temporal polysensory (STP) area, located on the upper bank of the superior temporal sulcus adjacent to area TE; 2) area PG in the parietal cortex; and 3) area TF, located in the posterior parahippocampal gyrus. Because all of these areas could conceivably have contributed to the spared visual recognition memory after early TE damage, we prepared a second group of monkeys with neonatal TE removal who, at 2 years of age, were given additional, combined lesions of areas STP, PG, TF, and TEO. A third group with neonatal TE removal was given, at 2 years of age, combined lesions of STP, PG, and TF, but not TEO. While these latter secondary lesions, like the ones limited to TEO, left recognition memory intact, the more extensive lesion, which included all four areas (PG, STP, TF, and TEO) produced a memory loss of the same magnitude as that seen after area TE lesions in adult monkeys. By contrast, combined lesions of areas STP, PG, TF, and TEO in adults that were previously normal produced only a relatively mild impairment, supporting the idea that these areas are not normally critical in adulthood for recognition memory. The results suggest that, during infancy, visual memory functions are widely distributed throughout many visual association areas but that, with further maturation, the function gradually becomes more localized to area TE. As a result of early damage to area TE, however, it appears that visual recognition memory processes remain widely distributed throughout the visual cortical areas, all of which, consequently, contribute to the sparing of visual memory function.

These studies have thus revealed a high degree of plasticity in the postnatal primate brain as a result of early brain damage. Although the TEO-amygdalar rewiring may not be relevant to the sparing of visual memory, it nonetheless suggests the potential for a remarkable degree of reorganization in the postnatal period. This neural reorganization further suggests that connections may be modified, not only by lesions, but in even more subtle ways that could have profound effects on the behavior of the individual. In addition, the behavior/lesion studies indicate that many areas of the cortex, perhaps by virtue of their diffuse anatomical connections, have the potential for multiple functions early in development, but that with maturation the functions gradually become localized within specific areas. An early brain lesion may cause a function to remain widely distributed and thereby account for the preservation of function that is often observed after early brain damage (*e.g.* see refs. *9, 12, 14, 28*). By studying the organization of the neural system that underlies cognitive and socioemotional behaviors, we can begin to correlate changes in the anatomy and neurochemistry with changes in behavior both aberrant and compensatory.

SUMMARY

Visual recognition memory ability does not reach adult levels in the monkey until 2 years of age. To examine the possibility that immaturity of one or more of the neural stations in the pathway for recognition memory is responsible for the late development of this function, we examined the effects of early damage to either the limbic structures, known to be critical for the formation of new memories, or of area TE, which may be critical for the long-term storage of visual memories. Assessment of the effects of these lesions on recognition memory indicates that compensatory mechanisms operate early to promote permanent recovery from neonatal TE but not from neonatal limbic lesions, suggesting that association areas of the cortex are immature at birth and possess greater plasticity than limbic structures. Our findings indicate that the maintenance of exuberant projections, together with reorganization in other cortical association areas mediate the sparing of memory functions after neonatal removal of area TE. In addition to the severe memory impairment produced by the neonatal limbic lesion, the animals also display severe behavioral abnormalities such as lack of social interactions, blank facial expressions, and motor stereotypies. The developmental time course and nature of these disturbances resemble those seen in autistic children.

Acknowledgment
We thank Mortimer Mishkin for his generous support during all phases of this work.

REFERENCES

1 Bachevalier, J. (1990). *Ann. N.Y. Acad. Sci.* **608**, 457–484.
2 Bachevalier, J. (1991). In *Advances in Neuropsychiatry and Psychopharmacology*, ed. Tamminga, C.A. and Schulz, S.C., vol. 1, pp. 129–140. New York: Raven Press.
3 Bachevalier, J. (1994). *Neuropsychologia* **32**, 627–648.
4 Bachevalier, J. and Merjanian, P.M. (1994). In *The Neurobiology of Autism*, ed. Bauman, M.L. and Kemper, T.L., pp. 146–169. Baltimore: Johns Hopkins Press.
5 Bachevalier, J. and Mishkin, M. (1994). *J. Neurosci.* **14**, 2128–2139.
6 Bachevalier, J., Brickson, M., and Hagger, C. (1993). *NeuroReport* **4**, 77–80.
7 Bachevalier, J., Hagger C., and Mishkin, M. (1991). In *Brain Work and Mental Activity, Alfred Benzon Symposium 31*, ed. Lassen, N.A., Ingvar, D.H., Raichle, M.E., and Friberg, L., pp. 231–240. Copenhagen: Munksgaard.
8 Bachevalier, J., Ungerleider, L.G., O'Neill, J.B., and Friedman, D.P. (1986). *Dev. Brain Res.* **25**, 302–308.
9 Bates, E. (1992). *Curr. Opin. Neurobiol.* **2**, 180–185.
10 Bauman, M.L. and Kemper, T.L. (1994). In *The Neurobiology of Autism*, ed. Bauman, M.L. and Kemper, T.L., pp. 119–145. Baltimore: Johns Hopkins Press.
11 Beauregard, M., Malkova, L., and Bachevalier, J. (1992). *J. Neurosci. Abstr.* **18**, 872.
12 Goldman-Rakic, P.S., Isseroff, A., Schwartz, M.L., and Bugbee, N.M. (1983). In *Handbook of Cognitive Development: Biology and Infancy Development*, ed. Mussen, P., pp. 281–344. New York: John Wiley.
13 Kling, A.S. and Brothers, L.A. (1992). In *The Amygdala*, ed. Aggleton, J.P., pp. 353–377. New York: Wiley-Liss, Inc.
14 Kolb, B. and Whishaw, I.Q. (1989). *Prog. Neurobiol.* **32**, 235–276.
15 LeDoux, J.E. (1992). In *The Amygdala*, ed. Aggleton, J.P., pp. 339–351. New York: Wiley-Liss, Inc.

16 Mahut, H. and Moss, M. (1986). In *The Hippocampus*, ed. Isaacson, R.L. and Pribram, K.H., vol. 4, pp. 241–280. New York: Plenum Press.
17 Malkova, L. Bachevalier, J., Kirkpatrick, B., Merjanian, P.M., and Mishkin, M. (1991). *IBRO Abstr.* **3**, 321.
18 Malkova, L., Bachevalier, J., Kirkpatrick, B., Merjanian, P.M., and Mishkin, M. (1991). *Int. Soc. Dev. Psychobiol., Abstr.* p. 39.
19 Malkova, L., Bachevalier, J., and Mishkin, M. (1993). *Soc. Neurosci. Abstr.* **19**, 372.
20 Malkova, L., Mishkin, M., and Bachevalier, J. (1995). *Behav. Neurosci.*, in press.
21 Merjanian, P.M., Bachevalier, J., Pettigrew, K.D., and Mishkin, M. (1989). *Soc. Neurosci. Abstr.* **15**, 302.
22 Meunier, M., Bachevalier, J., Mishkin, M., and Murray, E.A. (1993). *J. Neurosci.* **13**, 5418–5432.
23 Meunier, M., Bachevalier, J., Murray, E.A., Merjanian, P.M., and Richardson, R. (1991). *Soc. Neurosci. Abstr.* **17**, 337.
24 Mishkin, M. and Murray, E.A. (1994). *Curr. Opin. Neurobiol.* **4**, 200–206.
25 Mishkin, M. and Petri, H.L. (1984). In *Neuropsychology of Memory*, ed. Squire, L.R. and Butters, N., pp. 287–296. New York: Guilford Press.
26 Mishkin, M., Malamut, B., and Bachevalier, J. (1984). In *Neurobiology of Learning and Memory*, ed. Lynch, G., McGaugh, J.L., and Weinberger, N.M., pp. 65–77. New York: Guilford Press.
27 O'Neill, J.B., Friedman, D.P., Bachevalier, J., and Ungerleider, L.G. (1986). *Soc. Neurosci. Abstr.* **12**, 809.
28 Stiles, J. and Thal, D. (1993). In *Brain Development and Cognition: A Reader*, ed. Johnson, M., pp. 643–665. Oxford: Blackwell Publ.
29 Squire, L.R. and Zola-Morgan, S. (1991). *Science* **253**, 1380–1386.
30 Wang, J., Aigner, T., and Mishkin, M. (1990). *Soc. Neurosci. Abstr.* **16**, 617.
31 Webster, M.J., Ungerleider, L.G., and Bachevalier, J. (1991). *J. Neurosci.* **11**, 1095–1116.
32 Webster, M.J., Ungerleider, L.G., and Bachevalier, J. (1991). *NeuroReport* **2**, 769–772.
33 Webster, M.J., Bachevalier, J., and Ungerleider, L.G. (1994). *J. Comp. Neurol.* **352**, 213–226.

2

INVOLVEMENT OF AMYGDALA AND SEPTO-HIPPOCAMPUS IN EMOTION

HISAO NISHIJO, TOSHIAKI KITA,* RYOI TAMURA, SATOSHI EIFUKU, AND TAKETOSHI ONO

*Departments of Physiology and *Japanese Oriental Medicine, Faculty of Medicine, Toyama Medical and Pharmaceutical University, Toyama 930-01, Japan*

The limbic system, including the amygdala (AM) and hippocampus (HF), has been implicated in a variety of emotional as well as mnemonic and cognitive processes (*4, 6, 19, 29, 39*). The anatomical studies suggest parallel sensory processing in the AM and the HF; both receive highly integrated sensory information from cortical sensory systems (*1, 2*), and the AM directly, and the HF via septal nuclei (SP) project to common target areas in the brainstem including the hypothalamus which is essential in various emotional responses (*2, 40*).

It has been reported that the emotional responses are highly dependent on cognitive factors (*13*). Extensive studies using a Pavlovian conditioning indicate that the animals can well localize a relevant stimulus in the environment, and display emotional responses to such stimulus (*i.e.*, conditioned emotional response). In higher mammals, development of the cortical sensory systems enables them to recognize a wide variety of the complex stimuli. In turn, higher mammals can display emotional responses to a wide variety of the complex stimuli such as "death mask" (*13*). Lesion studies suggest that the AM, which receives highly integrated information from sensory association cortices (*2*), is involved in acquisition and expression of emotional responses to a specific stimulus (conditioned stimulus) associated with unconditioned stimulus (*18*, *35*), or in evaluation of the biological significance of such stimulus (*8*, *15*, *46*). This evidence suggests that the AM neurons in primates might respond to a variety of complex objects with biological significance.

On the other hand, the animals can display emotional responses to context without a conditioned stimulus when they are put in the same environment where they received a shock previously. Here, context is defined as various stimuli surrounding a conditioned stimulus. It has been reported that lesion of the HF suppressed emotional responses to context, but not to a conditioned stimulus (*16*, *35*). Similarly, the SP lesion did not suppress fear responses to a conditioned stimulus (*21*), nor did it suppress contact to a discrete stimulus related to an electric shock (*43*), although this lesion reduced the expression of fear- or anxiety-related behaviors in specific situations (contexts) (*34*). Since the SP-HF system is one of the important areas related to spatial coding (*23*, *29*), the spatial factor (place) could be a very important one in relation to context.

Our previous recording studies in monkeys indicated that the AM neuronal activity was topographically organized within the AM in terms of sensory modality (*25–27*, *44*). The affective contingency of the stimulus was one of the salient determinants for responsiveness of some AM neurons since in reversal or extinction test neuronal responses were readily suppressed (*24–27*, *44*). These findings suggest a role for AM in stimulus-affect association. Reward contingency of an object, however, was apparently not a crucial determinant for responsiveness of HF neurons (*28*, *41*), although responses of these neurons were context-dependent. Some HF neurons responded only when the monkey saw a rewarding object from a particular direction and in a particular place, or when the object was seen in a particular task (*7*, *28*, *30*, *31*). Responses of septal neurons might also be context-dependent since the SP is reciprocally connected to the HF (*40*), and the SP functions are closely correlated to those of the HF (*10*, *11*). However, there have been no studies on the role of septal nuclei in contextual coding. Recently, to investigate the role of the SP in emotion associated with contextual coding, activities of the monkey

SP neuron were analyzed during a task in which the biological significance of the object was dependent on the location of the monkey (*17*).

I. ANIMALS AND TASK PARADIGMS

Two male *Macaca fuscata* weighing 3.8 and 6.0 kg were used. The monkey was restrained painlessly in a stereotaxic apparatus by a surgically fixed head holder and sat in a chair in a cab which was set in a 2.5×2.5 m field located within a 5×6 m experimental room (Fig. 1A), as previously described (*30, 31*). During a recording session eye movements were monitored by electrooculograms (EOG). The room contained several visual landmarks, such as a stereomicroscope, a refrigerator, a table, a rack, and some experimental devices, and the monkey faced toward the right side of Fig. 1B. The cab was moved horizontally from one place to another by computer controlled electric motors. The front, left and right walls of the cab were made of half-mirrored Plexiglas and the rear wall was a steel plate equipped with two lamps (cab lamps) and two speakers on the inside. The landmarks in the room could be seen from inside of the cab through the half-mirrored walls only when the cab lamps were extinguished. The upper part of the front wall also contained a double half-mirrored 15×15 cm shutter just above an operant responding lever. There was an illuminated stage for object presentation behind the shutter and an object on the stage was visible to the monkey through the shutter when a front lamp was on. Juice, controlled by an electromagnetic valve, was delivered from a tube that projected through the rear wall. In this setup, monkeys were trained in the following three tasks: *Place-dependent go-nogo (PGN), Place-independent asymmetrical go-nogo (AGN), and Place-independent symmetrical go-nogo (SGN) tasks.*

In the PGN task, the cab was always placed in one of the four corners of the field (place I, II, III, or IV). One 3-dimensional object, selected from a total of four was presented, at one time or another, in each place. Each of the objects, which were rectangular parallelepipeds made of differently colored blocks, predicted a reward only when it was presented in one corresponding place (go trial), and the

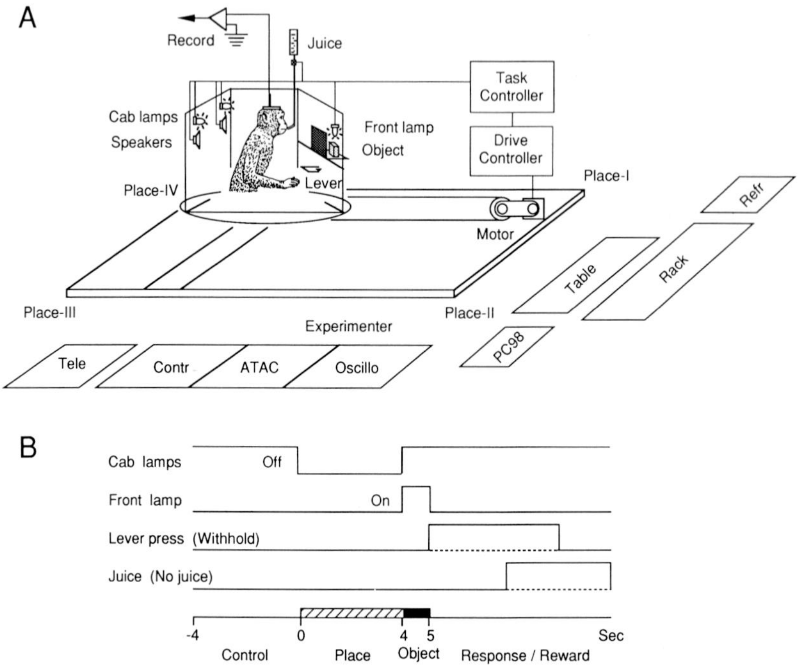

Fig. 1. Schema of the experimental setup and paradigm. A. Movable monkey cab in experimental shielded room (5×6 m). The monkey sat in a chair placed in a 0.7 m long×0.8 m wide×0.85 m high cab. The cab was moved to one of four places (Place-I, II, III, or IV) in the 2.5×2.5 m space. Front, left and right walls of the cab were half-mirrored. The rear wall, made of steel, was equipped with two cab lamps and speakers. A half-mirrored shutter (15×15 cm) was in the upper half of the cab front wall just above an operant lever. A front lamp behind the shutter illuminated an object in the bay. Tele, telemeter; Contr, controller; ATAC, microcomputer ATAC 3700; Oscillo, oscilloscope; PC98, microcomputer; Refr, refrigerator. B. Time sequences of PGN task. During 4 sec cab-lamp-off, the monkey could see the room environment through half-mirrored walls (place phase). While the front lamp was on for 1 sec, a given object was visible behind the shutter (object phase). If the object meant a reward, a drop of juice was dispensed from a spout after the lever was depressed for 3 sec (go trial). If the object meant a nonreward, juice was never available and pressing the lever produced a loud coarse buzzer sound (nogo trial).

same object predicted nonreward in all other places (nogo trials). For example, object 1 meant a reward in place I, and nonreward in places II, III or IV, and object 2 meant a reward only in place II, and so on. Consequently, each object presented two variable response/reinforcement contingencies depending on the location in which the monkey encountered it; there was a go/reward contingency and a nogo/nonreward contingency.

Each trial consisted of 4 phases (Fig. 1B). The control phase was a period of 4 sec before each trial. A trial was initiated by turning off the inside cab lamps with a click sound. While the cab lamps were off for 4 sec (place phase), the monkey could see outside of the cab through the half-mirrored cab walls. In the following object phase, a front lamp was lit for 1 sec to illuminate an object behind the half-mirrored shutter. If a given object meant a reward in that place (go trial), the monkey was required to initiate a response within 7 sec after the end of the object phase and hold the lever down for 3 sec. After the lever had been pressed for 3 sec, approximately 0.3 ml of an orange juice reward was presented for 4 sec. If a given object meant a nonreward in that place (nogo trial) and the monkey pressed the lever during the 7 sec period after the object phase, a buzzer tone was sounded for 1 sec to indicate a nonreward (an error nogo trial). If the lever had not been pressed for the 7 sec period in either a go or a nogo trial, the trial finished without any event (an error go trial and correct nogo trial). The cab was moved to the next place after the end of each trial during an intertrial interval (ITI) of 30 to 40 sec and arrived before the control phase of the next trial. In training, the same nogo trial was repeated when the monkey pressed the lever in a nogo trial so that the animal could learn to withhold an improper go response. Moreover, errors were punished by prolonging the ITI to a 1-min time-out both in the nogo trial and in the go trial.

In the AGN task, the cab was placed in any one of the four locations utilized in the PGN task, and two objects other than those used in the PGN task were presented at random. Regardless of the animal's location, a white object was always associated with a reward, and a yellow object indicated no reward. Except for this point, the other task sequences, such as presentation of the outside views (place phase), presentation of the objects (object phase), lever pressing in a go trial, and delivery of juice reward only in a go trial (response/reward phase), were the same as those of the PGN task. Since the object-response/reward contingency was constant in this task, this task was regarded as an unconditional place-independent version of the PGN task.

The SGN task was similar to the AGN task except that correct response in a nogo trial was associated with juice reward as in a go trial. The white object which was used in the AGN task was also presented in a go trial and a black object used only in this task was presented in a nogo trial. The white and black objects differed in response contingency, but were the same in reward contingency. That is, the black object was constantly associated with a nogo/reward contingency.

II. MONKEY SEPTAL NEURONAL RESPONSES

The activity of 430 neurons was recorded from the septal nuclei in two monkeys while they performed the PGN task. Of these neurons, 14 responded in the place phase but not in the object phase (place-phase neurons), 47 responded in the object phase but not in the place phase (object-phase neurons), and 129 responded in both place and object phases (both-phase neurons). In the present paper, we analyzed the responses of these 190 visually responsive neurons during the place and/or object phases.

1. Place-differential Responses in the Place Phase

Responses during the place phase were significantly place-differential in 58 (13 place-phase and 45 both-phase) neurons. Responses of the remaining 85 neurons were nondifferential. Response of a typical place-differential neuron is shown in Fig. 2A. This neuron responded during the place phase only when the cab was located in place I, but not during the object phase in the PGN task (Aa). The mean magnitudes of the responses during the place phase are shown in Fig. 2Ab. There were no differences in the spontaneous firing rates during the control phase among the 4 cab locations, and this was true of all the other place-differential neurons. This place-differential response was independent of a specific distal cue, since the responses were not affected by the environmental manipulation when the monkey learned the new environment in which a table in front of place II was moved in front of place I (not shown).

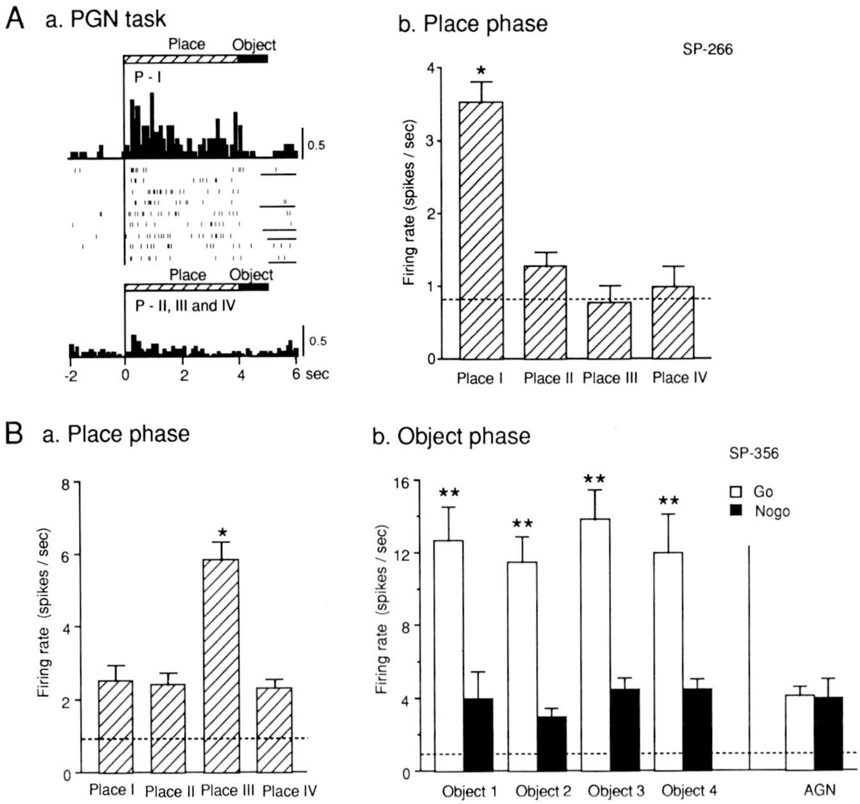

Fig. 2. Activities of a place-phase neuron with place-differential response in the place phase (A), and a both-phase neuron with place-differential response in the place phase and PGN-specific go/reward-selective response in the object place (B). A. Neuron was activated to place-I in the PGN task. a: comparison of place-phase neural responses in place I and in other places shown as a raster display and summed histograms (100 msec bins, 8 sec), regardless of the kinds of objects. There were no statistical differences among the place-phase responses in places II, III, and IV; therefore, these responses were included together in a summed histogram. Place phase (4 sec) is marked by horizontal hatched bar and object phase (1 sec) by filled bar. Lines under each raster mean lever press. The scale at the right of the histogram represents a mean firing rate (spikes/bin). b: histograms of mean firing rate ($\pm$SEM) during place phase in each animal's location, in which significant difference was revealed by ANOVA ($*p<0.001$). B. Neuron was activated when objects indicated go/reward contingency in the PGN task but not in the AGN task. a: place-phase responses (mean$\pm$SEM) for 4 locations, in which significant difference was revealed by ANOVA ($*p<0.001$). b: object-phase responses (mean$\pm$ SEM) to sight of each object are shown for go trials and nogo trials regardless of animal's location in the PGN (Objects 1–4) and AGN tasks. Significant difference between responses for go and nogo trials was revealed by Mann-Whitney U-test ($**p<0.01$). Broken line shows mean spontaneous firing rate.

2. PGN-specific Responses in the Object Phase

For each of the object- and both-phase neurons with object-phase responses ($n=176$), responses to the 4 objects in the PGN task were individually analyzed and compared with respect to the response/reinforcement contingency. Of these 176 neurons, 91 (34 of 47 object-phase; 57 of 129 both-phase) exhibited contingency-differential (*i.e.*, go/reward- and nogo/nonreward-selective) responses to one or more objects. Of 91 contingency-differential neurons, 53 go/reward-selective neurons responded more in go trials, and 38 nogo/nonreward-selective neurons responded more in nogo trials. A typical go/reward-selective neuron is shown in Fig. 2B. In the place-phase, this neuron had a place-differential response with preference for place III (Ba). Responses of this neuron were stronger to each object in go trials than in nogo trials (Bb).

To test whether or not contingency-differential response during the object phase in the PGN task depended on conditional place-object relations, 72 neurons (37 of 53 go/reward-selective; 35 of 38 nogo/nonreward-selective) were tested in the unconditional place-independent AGN task. Of these 72 neurons, 22 go/reward-selective and 21 nogo/nonreward-selective neurons responded differentially only in the PGN task in which discrimination of conditional place-object relations was required (PGN-specific). Typical examples are shown in Fig. 2Bb. The go/reward-selective neuron did not respond differentially to a go/reward object (white object) in the AGN task. The remaining 29 neurons (15 go/reward-selective; 14 nogo/nonreward-selective) were contingency-differential responsive in both the AGN and the PGN tasks (PGN-nonspecific).

3. Reward or Behavioral Contingency-differential Responses in the Object Phase

Of the 15 PGN-nonspecific go/reward-selective neurons, 9 were further tested in the SGN task, in which go and nogo responses were both reinforced to test whether this contingency-differential response reflected behavior (go response) or reward availability. Figure 3A and B shows the results for two different go/reward-selective neurons. Both responded strongly during the object phase when the objects

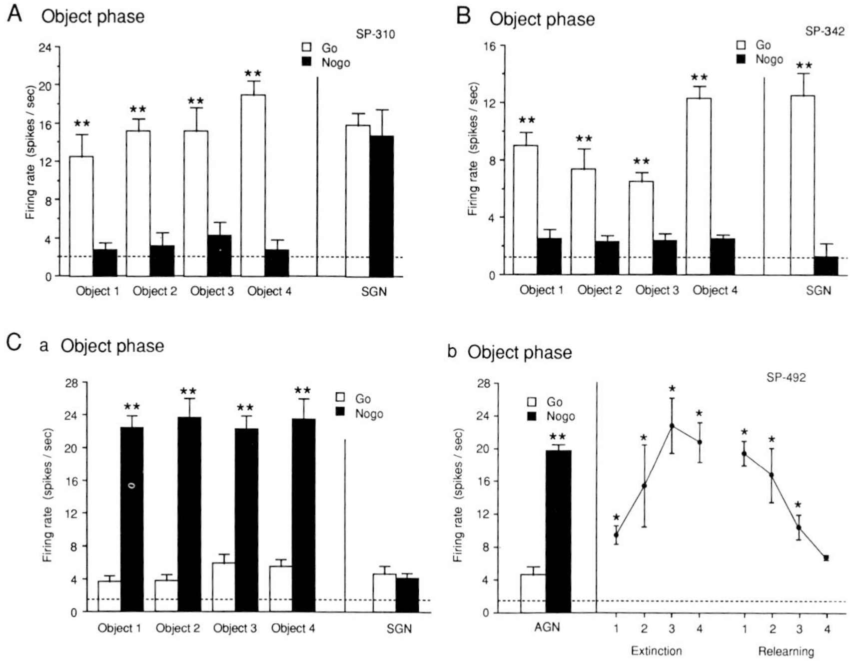

Fig. 3. Activity of a both-phase neuron with place-nondifferential response in the place phase (not shown) and reward-related response in the object phase (A), an object-phase neuron with go-related response in the object phase (B), and an object-phase neuron with nonreward-related response in the object phase (C). A. Neuron was activated when objects indicated reward contingency in the PGN and SGN tasks. B. Neuron was activated when objects indicated go contingency in both tasks. C. Neuron was activated when objects indicated nonreward contingency in PGN task (Object 1–4) and did not respond to the objects associated with reward regardless of go and nogo trials in SGN task. a,b: preferential response to the nonrewarding objects in the PGN, SGN (a), and AGN tasks (b). In Cb, the change of the firing rate at the sight of the white object is shown in extinction and relearning trials (mean and SEM of three trials). Note that responses to the rewarding white object increased after several nonreinforced extinction-trials and slowly returned after relearning. Other descriptions as for Fig. 2.

were rewarding, depending on their place-object relations in the PGN task (A and B). The neuron in Fig. 3A responded to the object indicating nogo/reward contingency as well as the object indicating go/reward contingency in the SGN task (reward-related neuron). On the contrary, the neuron in Fig. 3B did not respond to the object indicating nogo/reward contingency, but responded to the object indicating go/reward contingency in the SGN task (go-related neuron). Of the 9 go/reward-selective neurons that could be tested on the SGN task, 7 were reward-related, and 2 were go-related.

Of the 14 PGN-nonspecific nogo/nonreward-selective neurons, 12 were similarly tested with the SGN task. All of these 12 neurons were concluded to be nonreward-related neurons, since they did not respond to the object indicating nogo/reward contingency in the SGN task. A typical nonreward-related neuron is shown in Fig. 3C. This neuron responded to the object when it indicated nonreward contingency with nogo response, but did not respond when it indicated reward contingency with go response in the PGN (Ca) and AGN (Cb) tasks. In the SGN task, the neuron did not respond to the 2 objects indicating reward contingency with go or nogo response (Ca).

To characterize the reward/nonreward contingency related responses mentioned above, effects of direct manipulation of reward contingency were analyzed. Of the 12 nonreward-related neurons, 4 were further tested with extinction and reassociation of reward. Among these 4, plastic change of neural response was demonstrated by 2 neurons, whereas the responses of the other 2 neurons were not affected by extinction. The neuron depicted in Fig. 3C is an example of such neurons with plastic change. Activity of this neuron during presentation of the white object (usually rewarding) in the AGN task was analyzed during the course of extinction and subsequent relearning. Before extinction trials, the neuron responded to the yellow object indicating nonreward contingency, but not to the white object indicating reward contingency. Mean response magnitudes in the block of 3 trials during presentation of white object are shown in Fig. 3Cb in comparison to the responses in the ordinary AGN task. After several nonreinforced extinction trials with the white object, the monkey learned to withhold the lever press. The response during presentation of the white object became as strong as that during presentation of the nonrewarding yellow object. During the relearning trials, this response gradually decreased to the pre-test level.

Of the 7 reward-related neurons, 3 were similarly tested with extinction and relearning. The responses of all these 3 to the rewarding white object in the AGN task disappeared after extinction and quickly reappeared after relearning (not shown).

Thus, most of the PGN-nonspecific neurons (19/21, 90.5%)

which could be tested in the SGN task were related to the reward/ nonreward contingency, rather than to the response (behavior) contingency.

4. Location of Each Neuron Type

Many of the contingency-differential neurons (PGN-specific, reward-related, go-related, and nonreward-related) were densely packed in the dorsal and intermediate parts of lateral septal nuclei and some PGN-specific neurons were in the medial septal nuclei. Most neurons with place-differential responses were located in the area dorsal to the anterior commissure in the middle planes.

III. DUAL CONTROL SYSTEMS FOR EMOTIONAL EXPRESSION

Generally, the AM is believed to participate in the process through which sensory stimuli gain motivational and emotional significance (*8, 15, 38*). Previous recording studies in monkey AM (*25–28, 32, 33*) demonstrated that AM neuronal responses to various objects were related to their affective (rewarding or aversive) significance. The present study also revealed that septal neurons can differentiate reward availability.

Several functional differences between these two structures are also evident. In case of reward/nonreward-related SP neurons, some were correlated to reward contingency and others to nonreward contingency. On the other hand, few AM neurons responded to familiar meaningless objects (*25–28, 32, 33, 37*). Therefore, processing of nonreinforced events is likely to be related to the SP rather than to the AM. As hypothesized by Cormier (*3*), the SP seems to restrain the action of the AM and to damp reinforcing properties of stimuli associated with reinforcement for control of organismic responsiveness. A similar idea was proposed by Pribram (*36*), who suggested that the SP-HF system might be involved in active ignorance or the filtering out of a nonreinforced cue. Latent inhibition, which indicates states of decreased response to a conditioned stimulus as a result of prior nonreinforced experience, might be a behavioral manifestation of this function. This latent inhibition was disrupted by manipulations of the SP-HF system, but not of the AM (*9, 42, 45*).

This confirms the crucial role of the former system in the processing of nonreinforcement. Furthermore, latent inhibition has been reported to be context-specific (*12*) and HF lesions abolished this context specificity (*14*). Accordingly, the present contingency-differential response in the SP neurons also depended on spatial context.

These findings suggest that, while the AM might be essential for evaluating the affective property of a significant object itself, as suggested previously (*26–28*), the SP is involved in the integration of information from the environment surrounding the object, the affective significance of which varies based on the environmental situation. This functional difference may account for the results in previous behavioral studies in which the effects of AM and SP-HF lesions were sometimes opposite (*3*, *22*). Neurophysiological studies have demonstrated that the SP and the AM could both influence the same neuron in the hypothalamus (*5*), or in the ventral tegmental area (*20*). Taken together, these two regions could function as complements in motivational and emotional behaviors through projections to the same targets in the brainstem.

SUMMARY

The limbic system, including the AM and HF, has been implicated in a variety of emotional as well as mnemonic and cognitive processes. Both the AM and the HF receive highly integrated sensory information from cortical sensory systems, and the AM directly, and the HF via SP project to the hypothalamus which is essential in emotional behavior. Our previous recording studies in monkeys indicated that the affective contingency of the object was one of the salient determinants for responsiveness of some AM neurons, and suggest a role of AM in stimulus-affect association. On the other hand, reward contingency of an object was apparently not a crucial determinant for responsiveness of SP neurons. About 60% of the responsive SP neurons responded differentially to the objects only in the place-dependent conditional task, but not in the place-independent task. Most responses of the remaining neurons were related to the reward/nonreward contingency regardless of the place, but not to the behavioral contingency. The results suggest that septal nuclei are involved in integrating spatial information and reward/nonreward contingency.

On the basis of different functions of the AM and the SP-HF, these structures might function complementarily. These characteristics of AM and SP-HF neurons might be neurophysiological bases of human emotional and cognitive disorders.

Acknowledgments

This work was supported partly by the Japanese Ministry of Education, Science and Culture Grants-in-Aid for Scientific Research (05267103, 06454706, 06680786, 07244207, and 07558110), the Human Frontier Science Program for the 3rd fiscal year, and the Uehara Memorial Foundation.

REFERENCES

1 Amaral, D.G. and Insausti, R. (1990). In *The Human Nervous System*, ed. Paxinos, G., pp. 711–755. San Diego, CA: Academic Press.
2 Amaral, D.G., Price, J.L., Pitkänen, A., and Carmichael, S.T. (1992). In *The Amygdala*, ed. Aggleton, J.P., pp. 1–66. New York: John Wiley & Sons, Inc.
3 Cormier, S.M. (1981). *Physiol. Psychol.* **9**, 3–36.
4 Davis, M. (1994). *Int. Rev. Neurobiol.* **36**, 225–266.
5 Dreifuss, J.J. and Murphy, J.T. (1968). *Brain Res.* **8**, 167–176.
6 Eichenbaum, H., Otto, T., and Cohen, N.J. (1994). *Behav. Brain Sci.* **17**, 449–518.
7 Eifuku, S., Nishijo, H., Kita, T., and Ono, T. (1995). *J. Neurosci.*, in press.
8 Gaffan, E.A., Gaffan, D., and Harrison, S. (1988). *J. Neurosci.* **8**, 3144–3150.
9 Gallagher, M., Meagher, M.W., and Bostock, E. (1987). *Behav. Neurosci.* **101**, 315–324.
10 Gray, J.A. (1987). *The Neuropsychology of Anxiety*, New York: Oxford University Press.
11 Gray, J.A. and McNaughton, N. (1983). *Neurosci. Behav. Rev.* **7**, 119–188.
12 Hall, G. and Channell, S. (1985). *J. Exp. Psychol.* **11**, 470–481.
13 Hebb, D.O. (1972). *A Textbook of Psychology (3rd edition)*, Philadelphia: Saunders.
14 Honey, R.C. and Good, M. (1993). *Behav. Neurosci.* **107**, 23–33.
15 Jones, B. and Mishkin, M. (1972). *Exp. Neurol.* **36**, 362–377.
16 Kim, J.J. and Fanselow, M.S. (1992). *Science* **256**, 675–677.
17 Kita, T., Nishijo, H., Eifuku, S., Terasawa, K., and Ono, T. (1995). *J. Neurosci.* **15**, 1683–1703.
18 LeDoux, J.E. (1987). In *Handbook of Physiology, 1: The Nervous System*, ed. Mountcastle, V.B., vol. 5, pp. 419–459. Bethesda: American Physiological

Society.
19 LeDoux, J.E. (1992). *Curr. Opn. Neurobiol.* **2**, 191–197.
20 Maeda, H. and Mogenson, G. (1981). *Neuroscience* **6**, 367–376.
21 Melia, K.R. and Davis, M. (1991). *Physiol. Behav.* **49**, 603–611.
22 Melia, K.R., Sananes, C.B., and Davis, M. (1991). *Physiol. Behav.* **51**, 175–180.
23 M'Harzi, M. and Jarrard, L.E. (1992). *Behav. Brain Res.* **49**, 159–165.
24 Muramoto, K., Ono, T., Nishijo, H., and Fukuda, M. (1993). *Neuroscience* **52**, 621–636.
25 Nishijo, H., Ono, T., Nakamura, K., Kawabata, M., and Yamatani, K. (1986). *Brain Res. Bull.* **17**, 847–854.
26 Nishijo, H., Ono, T., and Nishino, H. (1988a). *J. Neurosci.* **8**, 3556–3569.
27 Nishijo, H., Ono, T., and Nishino, H. (1988b). *J. Neurosci.* **8**, 3570–3583.
28 Nishijo, H., Ono, T., Tamura, R., and Nakamura, K. (1993). *Prog. Brain Res.* **95**, 339–357.
29 O'Keefe, J. and Nadel, L. (1978). *The Hippocampus as a Cognitive Map*, Oxford: Clarendon, UK.
30 Ono, T., Nakamura, K., Fukuda, M., and Tamura, R. (1991). *Neurosci. Lett.* **121**, 194–198.
31 Ono, T., Nakamura, K., Nishijo, H., and Eifuku, S. (1993). *J. Neurophysiol.* **70**, 1516–1529.
32 Ono, T. and Nishijo, H. (1992). In *The Amygdala*, ed. Aggleton, J.P., pp. 167–190. New York: Wiley-Liss.
33 Ono, T., Nishijo, H., and Uwano, T. (1995). In *Progress in Neurobiology*, ed. Kerkut, G.A. and Phillis, J.W. New York: Pergamon Press, in press.
34 Pesold, C. and Treit, D. (1992). *Physiol. Behav.* **52**, 37–47.
35 Phillips, R.G. and LeDoux, J.E. (1992). *Behav. Neurosci.* **106**, 274–285.
36 Pribram, K.H. (1986). In *The Hippocampus*, ed. Isaacson, R.L. and Pribram, K.H., pp. 329–370. New York: Plenum Press.
37 Rolls, E.T. (1992). In *The Amygdala*, ed. Aggleton, J.P., pp. 143–165. New York: Wiley-Liss.
38 Spiegler, B.J. and Mishkin, M. (1981). *Behav. Brain Res.* **3**, 303–317.
39 Squire, L.R. and Zola-Morgan, S. (1991). *Science* **253**, 1380–1385.
40 Swanson, L.W., Köhler, C., and Björklund, A. (1987). In *Handbook of Chemical Neuroanatomy*, ed. Björklund, A., Hökfelt, T., and Swanson, L.W., vol. 5, pp. 125–277. New York: Elsevier.
41 Tamura, R., Ono, T., Fukuda, M., and Nishijo, H. (1992). *Hippocampus* **2**, 287–306.
42 Toga, A.W. and Burton, H.A. (1979). *J. Neurosci. Res.* **4**, 215–224.
43 Treit, D., Pesold, C., and Rotzinger, S. (1993). *Behav. Neurosci.*, **107**, 770–785.
44 Uwano, T., Nishijo, H., Ono, T., and Tamura, K. (1995). *Neuroscience*, in press.
45 Weiner, I. (1990). *Psychol. Bull.* **108**, 442–461.
46 Weiskrantz, L. (1956). *J. Comp. Physiol. Psychol.* **49**, 381–391.

3

THE ROLE OF CENTRAL DOPAMINERGIC AND SEROTONERGIC SYSTEMS OF RATS IN CONDITIONED FEAR STRESS, AN ANIMAL MODEL OF ANXIETY

TAKESHI INOUE, KIYOSHI TSUCHIYA, SHINJI HASHIMOTO, TAKESHI IZUMI, AND TSUKASA KOYAMA

Department of Psychiatry, Hokkaido University School of Medicine, Sapporo 060, Japan

On the basis of the growing evidence of the key role played by serotonergic systems in mediating the treatment of anxiety, several serotonin1_A (5-HT_{1A}) agonists and selective serotonin reuptake inhibitors (SSRIs) have been developed as new anxiolytics or antipanic agents (*1*). Elucidating the mechanism of action of 5-HT-related anxiolytics is likely to lead to new insights in our understanding of anxiety and related disorders. The efficacy of 5-HT-related anxiolytics suggests that brain serotonergic systems are closely related to the feeling of anxiety or coping with stress. However, there are a few studies investigating the relationship between 5-HT neurotransmission and anxiety using animal models of anxiety, and there are also a few concerning the effects of conditioned fear stress (CFS) (*11*) on the serotonergic system.

This paper summarizes experimental findings in which we have identified i) effects of CFS on fear-related behavior (freezing), ii) stimulatory effects of CFS on 5-HT and dopamine (DA) metabolism in the rat brain (*17, 18*), and iii) anxiolytic effects of various 5-HT-related drugs on conditioned freezing, as an index of anxiety (*15, 19*).

I. EFFECT OF CONDITIONED FEAR STRESS ON BEHAVIOR AND PLASMA CORTICOSTERONE

CFS is regarded as an animal model of anxiety and psychological stress because It includes no physical stimuli, at least at the time of testing (*11*). The procedure of CFS is very simple and based on classical aversive conditioning: in brief, rats are individually subjected to footshock stress in a shock box, and again placed in the same shock box without shocks 24 hr later. In our experiment, the rats exposed to CFS exhibited increased defecation and freezing behavior (*17*); this latter was defined as the lack of all observable movement of the body and the vibrissae except that related to respiration (*10*). Fanselow suggested that the freezing response could be used as an index of activation of the fear system (*11*). Control rats, which were placed in the shock box without shocks and observed in the same shock box without shocks 24 hr later, did not show freezing at all. CFS also elevated plasma corticosterone concentrations up to 40–60 μg/dl, which was close to the ceiling of stress-induced elevation. These all indicate the validity of CFS as emotional stress.

CFS-induced freezing (conditioned freezing) is a good index of the stress intensity of CFS, compared with defecation or plasma corticosterone levels. As shown in Table I, we reported that conditioned freezing was enhanced by repeated conditioning and an increase in the intensity of footshock, which were assumed to augment conditioned fear (*18*). These results are consistent with previous reports (*10*, *12*).

TABLE I
Effects of CFS on Stress-induced Increases in Defecation, Plasma Corticosterone Concentrations and Freezing

	CFS	Effect of an increase in shock intensity	Effect of repeated conditioning
Defecation	↑	Unchanged	Unchanged
Plasma corticosterone	↑	n.d.	n.d.
Freezing	↑	Enhanced	Enhanced

n.d., not determined. From refs. *17* and *18*, and our unpublished data.

II. EFFECT OF CONDITIONED FEAR STRESS ON DA AND 5-HT METABOLISM

CFS has been reported to selectively increase the DA metabolism in the medial prefrontal cortex (mPFC) (*8, 16*), and also to increase the noradrenaline (NA) metabolism in the amygdala, hypothalamus and locus coeruleus (*27*). Recent clinical studies have suggested that the brain serotonergic system plays an important role in anxiety (*9*), but there are few studies on the relationship between brain 5-HT metabolism and CFS. To elucidate the 5-HT function during CFS, we examined the effects of CFS on 5-HT and DA metabolism in the rat brain (*17, 18*). Monoamine concentrations (DA, 5-HT, and their metabolites) were measured using high performance liquid chromatography with electrochemical detection in seven brain regions of male Sprague-Dawley rats (250–300 g) exposed to CFS after single or repeated footshock sessions (Table II).

As shown in Table II, 2.5 mA of electric footshock stress for 30 min increased the DA metabolism in most brain regions examined, and increased the 5-HT metabolism in the mPFC and two other brain regions. Although an increased metabolism of 5-HT and DA is not always indicative of increased neuronal activity (*5*), we con-

TABLE II
Effects of CFS after Single and Repeated Footshock Sessions (Conditioning) on Brain Dopamine and Serotonin Metabolism

	Dopamine			Serotonin		
	EFS	sCFS	rCFS	EFS	sCFS	rCFS
Medial prefrontal cortex	↑	↑	↑	↑	↑	↑
Nucleus accumbens	↑		↑	↑		↑
Striatum	↑					
PVH	↑	↑	↑			
Amygdala	↑		↑			↑
Lateral hypothalamus	↑	↑	↑	↑		
Hippocampus			↑			

From ref. *18*. PVH, paraventricular nucleus of the hypothalamus; EFS, electric footshock stress; sCFS, CFS after single footshock; rCFS, CFS after repeated footshock. ↑, increase; blank, no change.

firmed, using *in vivo* microdialysis method, that the increased metabolism of 5-HT and DA in the mPFC during stress reflects an increase in the release of synaptic 5-HT and DA (our unpublished data). In contrast to footshock stress, CFS for 30 min following a single footshock session (2.5 mA for 30 min) selectively increased the 5-HT metabolism in the mPFC, and increased the DA metabolism in the mPFC, paraventricular nucleus of the hypothalamus and lateral hypothalamus (*17, 18*). In summary, while physical stress (footshock) increased the DA and 5-HT metabolism in several brain regions, psychological stress (CFS) increased the DA and 5-HT metabolism only in the mPFC and a few other regions. Similar findings about the effect of psychological stress on brain DA metabolism have been reported by other investigators (*8, 16, 21*). They reported that psychological stress selectively increased the DA metabolism in the mPFC. However, it is not clear whether regional differences between footshock and CFS in DA and 5-HT activation are due to the intensity of stress or the quality of stress: one cannot obviate the possibility that DA and 5-HT activation during footshock in more widespread regions than CFS reflect not only emotional responses but also physical influences such as pain.

To resolve this question, we compared the effects of two intensities of CFS and footshock on DA and 5-HT metabolism in the rat brain (*18*). As mentioned in section I, rats that were repeatedly exposed to 2.5 mA footshock for a period of 10 days displayed a greater degree of freezing induced by CFS than those given only one footshock session, indicating an augmentation of fear and CFS intensity. CFS after repeated footshock increased the DA metabolism in most of the brain regions except for the striatum, and increased the 5-HT metabolism in the mPFC, nucleus accumbens and amygdala. These results suggest that regional patterns of brain DA and 5-HT activation during psychological stress depend on the intensity of that stress; and that the more widespread activation of DA and 5-HT after more severe CFS might be related to behavioral changes which reflect the augmentation of fear. Therefore, it is concluded that the regional differences between footshock and CFS after single conditioning in DA and 5-HT activation can be ascribed to the difference between two kinds of stress in the intensity of emotional responses to stress, although striatal DA metabolism was increased by physical stress, but not by either mild or severe psychological stress.

III. SEROTONERGIC ACTIVATION REDUCES DEFENSIVE FREEZING IN THE CONDITIONED FEAR PARADIGM

Although the above results of neurochemical experiments support the hypothesis that brain 5-HT is closely related to anxiety (*9*), the functional significance of the 5-HT response to stress is unclear. We investigated the effects of 5-HT reuptake inhibitors, agonists, antagonists, and diazepam on freezing behavior induced by CFS using a time-sampling procedure (*19*). Various doses of test compounds were administered subcutaneously to male Sprague-Dawley rats 24 hr after the last session of repeated footshock for 5 days. Rats were again placed in the shock chamber without shocks 20 min after injection of drugs and observed. As summarized in Table III, the classical anxiolytic diazepam (1 mg/kg) and the 5-HT_{1A} agonist ipsapirone (0.5–10 mg/kg) significantly inhibited conditioned freezing. The 5-HT precursor L-5-hydroxytryptophan (with peripheral amino acid decarboxylase inhibition by means of benserazide) and

TABLE III
Effects of Various Drugs on Conditioned Freezing

Drug		Response
Diazepam	Benzodiazepine	+
Ipsapirone	5-HT_{1A} agonist	+
L-5-HTP	5-HT precursor	+
PCPA	Tryptophan hydroxylase inhibitor	0
ICI169,369	5-$HT_{2/1C}$ antagonist	0
Citalopram	5-HT reuptake inhibitor	+
Fluvoxamine	5-HT reuptake inhibitor	+
ORG4428	NA reuptake inhibitor	0
GBR12909	DA reuptake inhibitor	0
LY288513	CCK_B antagonist	+

Rats were subjected to footshock in the shock box. Twenty-four hours after the last footshock session, drugs were injected to rats. Twenty to sixty min after the injections, the rats were again placed in the same shock box without shocks and observed (PCPA was administered 15 hr before the text). Freezing behavior was observed as an index of fear. Reduction in conditioned freezing was regarded as anxiolytic. From refs. *15*, *19*, and *20*. +, anxiolytic; 0, ineffective.

the SSRI citalopram (10 mg/kg) also reduced conditioned freezing. The 5-HT_2 antagonists ICI169,369 and ketanserin failed to change conditioned freezing. Tryptophan hydroxylase inhibitor *p*-chlorophenylalanine (200 mg/kg), administered 15 hr before the test, did not affect freezing. The effect of ipsapirone was not modified in rats with lesions of 5-HT neurons produced by *p*-chloroamphetamine (2×10 mg/kg). These results suggest the anxiolytic potential of ipsapirone and citalopram, and support the hypothesis that facilitation of 5-HT neurotransmission decreases anxiety (*9*).

Recent clinical studies have shown that L-5-hydroxytryptophan, 5-HT_{1A} agonists and SSRIs are effective in the treatment of anxiety disorders such as panic disorder and generalized anxiety disorder (*1*, *9*). Our data concerning the anxiolytic effects of these 5-HT-related drugs on CFS suggests that CFS is useful for the prediction of compounds effective in the treatment of anxiety disorders. In addition, the present results that drugs stimulating 5-HT receptor function attenuated conditioned freezing suggest that 5-HT activation during CFS shown in section II might be associated with coping with fear and anxiety. However, further studies are necessary to clarify the roles of serotonergic systems in various regions for anxiety.

IV. EFFECT OF 5-HT, NA AND DA REUPTAKE INHIBITORS ON CONDITIONED FREEZING

There is accumulating evidence from treatment response studies that antidepressants are therapeutically effective in anxiety disorders (*23*). Imipramine has been reported to alleviate symptoms of generalized anxiety disorder and panic disorder (*23*, *26*). It has been further reported that clomipramine, a potent 5-HT reuptake inhibitor, exerts anxiolytic activity (*23*). Interestingly, den Boer and Westenberg reported in 1988 that the SSRI fluvoxamine was effective in the treatment of panic disorder, while the NA reuptake inhibitor maprotiline was ineffective (*7*). These results indicate that 5-HT reuptake inhibition rather than NA reuptake inhibition is effective in the treatment of anxiety disorder, and the efficacy of imipramine and clomipramine is due to the 5-HT reuptake inhibition exerted by these compounds (*9*). Although anxiolytic activity of the SSRIs has been assessed in only a few animal models, such as suppressed feeding in a novel environment (*2*) and elevated plus-maze (*13*), some previous studies failed to demonstrate the anxiolytic effect of SSRIs and indicated that SSRIs were anxiogenic (*4, 13*). Furthermore, until now, there have been no studies that compared the effects of SSRIs and selective NA reuptake inhibitors in animal anxiety tests.

We compared the effects of SSRIs, NA reuptake inhibitors and DA reuptake inhibitors on conditioned freezing (*15*). In this experiment, male Sprague-Dawley rats were individually subjected to single footshock session for 5 min, and again placed in the same shock box without shocks 24 hr later and observed for 10 min. Drugs were injected to rats 60 min before testing. As shown in Table III, the SSRIs (citalopram and fluvoxamine) were effective in reducing conditioned freezing, but the selective NA reuptake inhibitor (ORG4428) and the selective DA reuptake inhibitor (GBR12909) were not effective. These results are consistent with clinical data that 5-HT, but not NA, reuptake inhibition is effective in reducing anxiety (*7*).

V. EFFECTS OF THE SELECTIVE CCK_B ANTAGONIST ON CONDITIONED FREEZING

Recent studies have indicated that brain cholecystokinin (CCK) has the anxiogenic potential in human and animals, and that selective CCK_B antagonists show anxiolytic effects in several standard anxiety tests (*14*). Although the anxiolytic effects of CCK_B antagonists have not been confirmed in anxiety disorders yet, a recent study demonstrated that the selective CCK_B antagonist L-365,260 antagonized the panicogenic effects of CCK-4 in patients with panic disorder (*3*). This suggests the possibility that CCK_B antagonists are anxiolytic in human. The anxiolytic effects of CCK_B antagonists in animals have been examined in elevated plus-maze test and social interaction test (*14*), but not in conditioned fear. As shown in Table III, we confirmed the anxiolytic effect of the selective nonpeptide CCK_B antagonist LY288513 in the conditioned fear test (*20*). Although CFS is closely associated with brain DA and 5-HT systems as described in section II, the effects of CCK_B antagonists on brain monoamine systems are not yet known. Future experiments will be necessary to examine the relationship between CCK_B receptors and brain monoamines, especially under stress.

VI. THE RELATIONSHIP BETWEEN BEHAVIORAL AND NEUROCHEMICAL CHANGES DURING CFS

How are the neurochemical changes during CFS related to behavioral changes such as freezing? As shown in section II, increased DA and 5-HT activation during CFS were especially marked in the mPFC compared with other brain regions. Moreover, it should be noted that CFS after single conditioning selectively increased the 5-HT metabolism in the mPFC. These results suggest that the mPFC may play an important role in CFS, although its functional significance during CFS has not been fully clarified.

We examined the effects of bilateral lesions of the mPFC on the acquisition of conditioned freezing. Bilateral thermal lesions of the mPFC were made under pentobarbital anesthesia with the electrode tip of a radiofrequency lesion generator (Radionics Inc., Burlington, MA) maintained at 60°C for 1 min. The following stereotaxic coordinates (*24*) were used: AP +3.0 mm from bregma, ventral −4.0 mm from the skull surface, and 1 mm from the midline. Male Sprague-Dawley rats were subjected to footshock for 20 min 2 weeks following lesions, and observed in the shock box without shocks for 10 min 24 hr after footshock. As shown in Fig. 1, the bilateral mPFC lesions did not reduce conditioned freezing at all. Therefore, it is concluded that dopaminergic and serotonergic activation in the mPFC during CFS do not produce freezing behavior.

Other investigators also reported that mPFC lesions did not reduce the acquisition of conditioned freezing, but that lesions of amygdala and hippocampus did reduce it (*22, 25*). What is the role of dopaminergic and serotonergic activation in the mPFC during CFS? Recently, Morgan *et al.* reported that mPFC lesions interfered with the extinction of conditioned freezing (*22*). Furthermore, D'Angio *et al.* suggested that the mPFC DA activation induced by stress is not connected to the emotional reaction caused by the aversive nature of the stressor, but may rather reflect the heightened attention of the animal or the activation of cognitive processes in an attempt to cope with the stressor (*6*). Taken together, although we failed to demonstrate it directly, these findings suggest that the mPFC

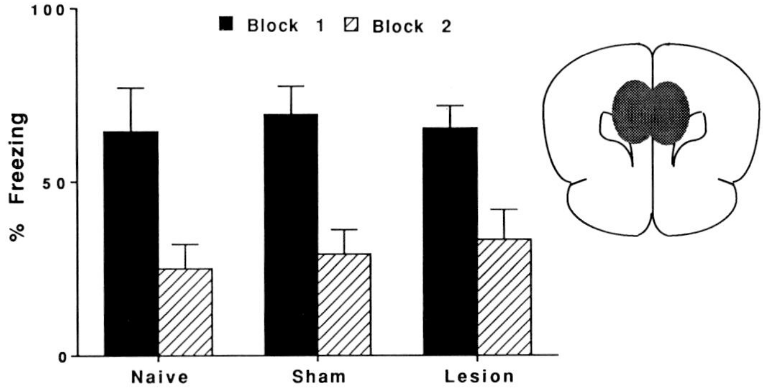

Fig. 1. Effect of bilateral lesions of the medial prefrontal cortex on the acquisition of conditioned freezing. The lesioned area is shown by stippling. Naive rats did not receive any surgery or anesthesia; sham-operated animals received the same treatment as the lesioned animals, but the electrode tip was not inserted; lesioned rats received radiofrequency lesions under pentobarbital anesthesia. All the rats received 20 min of footshock in the shock box 14 days after surgery, and were again placed in the same box without shocks 24 hr after footshock and observed for 10 min. Block 1, 0–5 min; Block 2, 5–10 min. $N = 16$ except for naive group ($N = 8$).

might play an important role in the coping or cognitive processes during fear and anxiety.

Finally, as discussed in section III, our results that drugs stimulating 5-HT receptor function attenuated conditioned freezing also suggest that 5-HT activation is associated with coping with fear and anxiety. Although 5-HT activation in the mPFC during CFS might be associated with coping, there is no evidence to support the functional significance of the mPFC 5-HT response to CFS. Further studies are necessary to clarify the roles of serotonergic systems in the mPFC and other brain regions for anxiety.

SUMMARY

The present review summarizes a series of neurochemical and behavioral studies on CFS, an animal model of anxiety. CFS increased defecation, elevated plasma corticosterone levels, and produced freezing behavior, an index of anxiety. CFS after single conditioning produced DA and 5-HT activation markedly in the mPFC. The augmentation of CFS intensity resulted in the more widespread activation of DA and 5-HT, which might be associated with the augmentation of anxiety. Various anxiolytics decreased conditioned freezing: these are diazepam, SSRIs, 5-HT_{1A} agonist, 5-HT precursor, and CCK_B antagonist. These results are consistent with clinical effects of these drugs in the treatment of anxiety disorders, and suggest that facilitation of 5-HT neurotransmission decreases anxiety because SSRIs and 5-HT precursor are assumed to increase the output from the 5-HT synapse. The bilateral mPFC lesions did not reduce conditioned freezing. This lesion study and other previous reports suggest that DA and 5-HT activation in the mPFC during CFS is not associated with emotional reaction to CFS, but may, rather, reflect coping with stress.

Acknowledgment
Our work quoted in this review was partly supported by research grants from the Ministry of Education, Science and Culture (Japan).

REFERENCES

1 Baron, R., Pohl, R., and Yeragani, V.K. (1990). In *Progress in Basic Clinical Pharmacology*, ed. Pohl, R. and Gershon, S., vol. 3, pp. 284–310. Basel: Karger.
2 Bodnoff, S.R., Suranyi-Cadotte, B., Quirion, R., and Meaney, M.J. (1989). *Psychopharmacology* **97**, 277–279.
3 Bradwejn, J., Koszycki, D., Couetoux du Tertre, A., van Megen, H., den Boer, J., Westenberg, H., and Annable, L. (1994). *Arch. Gen. Psychiatry* **51**, 486–493.
4 Chopin, P. and Briley, M. (1987). *Trends Pharmacol. Sci.* **8**, 383–387.
5 Commissiong, J.W. (1985). *Biochem. Pharmacol.* **34**, 1127–1131.
6 D'Angio, M., Serrano, A., Driscoll, P., and Scatton, B. (1988). *Brain Res.* **451**, 237–247.
7 Den Boer, J.A. and Westenberg, H.G. (1988). *Int. Clin. Psychopharmacol.* **3**, 59–74.
8 Deutch, A.Y., Tam, S.-Y., and Roth, R.H. (1985). *Brain Res.* **333**, 143–146.
9 Erikkson, E. and Humble, M. (1990). In *Progress in Basic Clinical Pharmacology*, ed. Pohl, R., and Gershon, S., vol. 3, pp. 66–119. Basel: Karger.
10 Fanselow, M.S. (1980). *Pavlov. J. Biol. Sci.* **15**, 177–182.
11 Fanselow, M.S. (1984). *Trends Neurosci.* **7**, 460–462.
12 Fanselow, M.S. and Helmstetter, F.J. (1988). *Behav. Neurosci.* **102**, 233–243.
13 Griebel, G., Moreau, J.L., Jenck, F., Misslin, R., and Martin, J.R. (1994). *Psychopharmacology* **113**, 463–470.
14 Harro, J., Vasar, E., and Bradwejn, J. (1993). *Trends Pharmacol. Sci.* **14**, 244–249.
15 Hashimoto, S., Inoue, T., and Koyama, T. (1994). *Jpn. J. Psychopharmacol.* **14**,

411 (in Japanese).
16 Herman, J.P., Guillonneau, D., Dantzer, R., Scatton, B., Semerdjian-Rouquier, L., and Le Moal, M. (1982). *Life Sci.* **30**, 2207–2214.
17 Inoue, T., Koyama, T., and Yamashita, I. (1993). *Pharmacol. Biochem. Behav.* **44**, 371–374.
18 Inoue, T., Tsuchiya, K., and Koyama, T. (1994). *Pharmacol. Biochem. Behav.* **49**, 911–920.
19 Inoue, T., Tsuchiya, K., and Koyama, T. (1995). *Pharmacol. Biochem. Behav.*, in press.
20 Izumi, T., Inoue, T., and Koyama, T. (1994). *Jpn. J. Psychopharmacol.* **14**, 462 (in Japanese).
21 Kaneyuki, H., Yokoo, H., Tsuda, A., Yoshida, M., Mizuki, Y., Yamada, M., and Tanaka, M. (1991). *Brain Res.* **557**, 154–161.
22 Morgan, M.A., Romanski, L.M., and LeDoux, J.E. (1993). *Neurosci. Lett.* **163**, 109–113.
23 Nutt, D.J. and Glue, P. (1989). *Pharmacol. Ther.* **44**, 309–334.
24 Paxinos, G. and Watson, D. (1982). *The Rat Brain in Stereotaxic Coordinates,* New York: Academic Press.
25 Phillips, R.G. and LeDoux, J.E. (1992). *Behav. Neurosci.* **106**, 274–285.
26 Rickels, K., Downing, R., Schweizer, E., and Hassman, H. (1993). *Arch. Gen. Psychiatry* **50**, 884–895.
27 Tsuda, A., Tanaka, M., Ida, Y., Tsujimaru, S., Ushijima, I., and Nagasaki, N. (1986). *Pharmacol. Biochem. Behav.* **24**, 115–119.

4

PET AND THE FUNCTIONAL ANATOMY OF MAJOR DEPRESSION

WAYNE C. DREVETS

Departments of Psychiatry and Radiology (Radiation Sciences), Washington University School of Medicine, St. Louis, MO 63110, U.S.A.

Functional imaging technologies such as positron emission tomography (PET) hold great promise for elucidating the pathophysiology of the major depressive disorders. These disorders are associated with a variety of neurochemical, neuroendocrine, and neuropsychological abnormalities, yet their biological bases have not been established. This is partly because the general absence of gross anatomical changes in mood disorders has obscured their pathophysiologic correlates. Nevertheless, the preservation of the underlying anatomy renders mood disorders particularly tractable to investigations involving functional brain imaging.

However, since physiological images reflect an averaging of numerous biochemical events that occur within the temporal resolution of the PET scan, abnormalities found in depressed samples are open to an imposing number of interpretations. Regional abnormalities may represent: 1) pathophysiologic changes associated with the genetic and/or environmental predisposition to depressive episodes, 2) state-dependent physiologic changes correlated with the cognitive, emotional or behavioral manifestations of the depressive syndrome, or 3) structural changes resulting from the illness or past treatment

(*9*, *42*). Since image data are affected by so many variables, thoughtful strategy is required so that successive experiments systematically isolate the possible causes of regional differences.

My colleagues and I have addressed this complexity using converging evidence from multiple types of imaging data. We have employed depression's episodic nature and response to treatment to obtain scans in the presence or absence of symptoms to permit differentiation of physiologic correlates of symptoms from pathophysiologic changes that may underlie the tendency to become depressed. We have also compared blood flow (BF) image data from depressed subjects with image data from normal subjects during experimentally induced sadness, anxiety, or negative ruminations—since many symptoms of major depression represent exaggerations or extensions of cognitive and emotional states experienced by non-depressed subjects. Finally, since the depressive disorders appear biologically heterogenous, we have contrasted image data across subtypes. Using depressed subjects from one subtype as control subjects for comparisons involving another subtype permits differentiation of abnormalities related to the pathophysiologic changes unique to that subtype from physiologic changes that are simply associated with depressive symptoms (which are present in both subtypes).

Subtyping depressed subjects has also proven critical for reducing the variability of data from depressed samples. The major depressive syndrome appears to be associated with multiple pathophysiologic states and is likely to be characterized by an assortment of distinct PET images as different subtypes are considered (*11*). Nevertheless, classification of depressed subjects a priori is problematic since subtypes cannot yet be sensitively or specifically identified using biological markers (*22*, *61*). Clinical and family history criteria must therefore be relied upon for obtaining relatively homogenous subject samples (*61*).

Two broad categories of primary major depression are recognized based upon clinical course ("primary" designates that the depressive syndrome antedated other medical or psychiatric illnesses that might be associated with it (*61*)): bipolar, in which both manic and depressive episodes occur, and unipolar, in which only depressive episodes occur. Unipolar and bipolar depressives appear neurochemically, pharmacologically and genetically distinct (*22, 34, 44, 48*). Consistent with these differences, unipolar and bipolar depressives differ in their respective circulatory and metabolic correlates (*3, 6, 11, 51*).

While additional etiological heterogeneity likely exists within both primary unipolar and bipolar disorder, in the absence of pathophysiology-based taxonomies we employ a classification for unipolar depression that capitalizes on the heritability of affective disorders, identifying subtypes based upon family history (*61*). Of these subtypes, familial pure depressive disease (FPDD—defined as primary unipolar depression in an individual who has a parent, sibling of offspring with primary unipolar depression but no family history of alcoholism, antisocial personality or mania) has been most consistently associated with biological markers (*61*). We thus selected subjects from three groups: 1) primary, unipolar depressives who meet criteria for FPDD, 2) unipolar depressives who fail to meet criteria for FPDD, and 3) bipolar depressives.

I. THE FUNCTIONAL ANATOMY OF FAMILIAL PURE DEPRESSIVE DISEASE

In an initial study of FPDD we scanned subjects in either the depressed or the remitted (asymptomatic) phase (*13*). The currently depressed subjects ($n = 13$) met DSM III-R criteria for Recurrent, Major Depression, and FPDD (mean age $= 36.2 \pm 8.9$ (range 22–52); 54% female, all right-handed) and had Hamilton Depression Rating Scale (HDRS) scores in the moderately to severely depressed range (mean score $= 27.3 \pm 4.6$ (range 21–35)). These subjects were unmediated for ≥ 3 weeks (mean drug-free period $= 7.5 \pm 9.6$ months). The remitted subjects ($n = 10$) met these criteria in the past (*i.e.*, they had suffered unipolar depressive episodes and met criteria for FPDD) but had been asymptomatic and unmedicated for ≥ 4 months prior to imaging (mean duration of remission $= 2.3 \pm 2.2$ yrs. (range 0.33 to 7 yrs.); mean age $= 33.6 \pm 10.0$ (24–55); 70% female, all right-handed). Subjects were clinically euthymic and had HDRS scores < 7. These groups were compared with a control group who denied having a personal or family history of psychiatric illness ($n = 33$; mean age $= 30.1 \pm 7.8$ (22–50); 61% female, all right-handed).

BF scans were acquired using 60–80 mCi of $H_2{}^{15}O$, PETT VI, and the eyes-closed, at-rest behavioral state. Images were reconstructed to a transaxial resolution of 17 mm full-width at half-maximum (FWHM). To negate effects of global fluctuations on regional activity, a linear normalization was applied (normalization reduces the standard deviation of paired BF measures obtained in the same scan session to 1%–2% and is thus critical to the sensitivity of PET studies involving small samples (Drevets and Spitznagel, unpublished data)).

Regional flow was increased in the left prefrontal cortex (PFC) and the left amygdala in FPDD (Fig. 1) (*13*). In the left PFC, BF was elevated in the depressed but not the remitted phase. In contrast, left amygdala flow appeared increased in both phases. These data suggested that FPDD is associated with an abnormality related to the state of being depressed in the left PFC, and possibly with an abnormality related to the trait of being susceptible to depressive episodes in the amygdala. Consistent with these hypotheses, we demonstrated that antidepressant drug treatment, which both ameliorates depressive symptoms and prevents recurrences of future

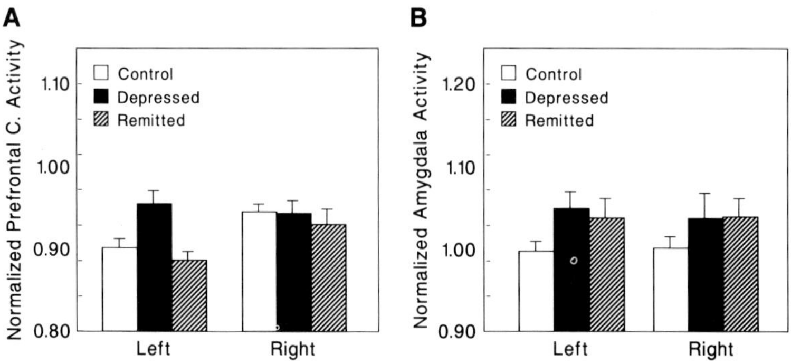

Fig. 1. A. Mean BF in the left and right PFC for control, FPDD-depressed and FPDD-remitted groups. In the left PFC the differences between the depressed and control groups, and between the depressed and remitted groups were significant. Differences between the left and right PFC were significant for the control and remitted groups, but not for the depressed group. B. Mean BF in the left and right amygdala. In the left amygdala, the difference between the depressed and control groups was significant. Values are mean normalized, tissue radioactivity, which accurately reflects normalized regional BF due to the linear relationship between flow and radioactive counts. Error bars represent SEM.

episodes (*26*), results in reduced BF and metabolism in the ventrolateral (VLPFC) and the amygdala bilaterally, compatible with preclinical data indicating that antidepressants suppress amygdala function (*10, 12, 59*).

In assessing the laterality of the abnormalities in the unmedicated subjects with FPDD we found that activity in the left PFC was less than in the right in the control (-5.0%, $t=4.1$, $p<.003$) and the remitted subjects (-5.0%, $t=2.6$, $p<.03$) but not the depressed subjects ($+1.8\%$, $t=0.97$, n.s., Fig. 1A). Mean activity in the right PFC did not differ among the three groups. Thus in the PFC, a normal right>left asymmetry was found in controls which was lost in the depressed phase, but restored in the remitted phase of FPDD.

In the left amygdala, activity did not differ from right in any of the groups (Fig. 1B). Mean activity in the right amygdala was 4.1% greater in the depressives relative to the controls, but this was not significant. The comparable 4.3% difference between the remitted and control groups in the right amygdala was also not significant ($t=1.6$).

The affected area in the left PFC included the VLPFC, the frontal polar cortex, the lateral and medial orbital cortices and the pregenual portion of the anterior cingulate gyrus (*13, 41*). The left VLPFC included the caudal part of Brodmann's area 11 (comparable to Walker's area 12), and areas 45 and 47. The frontal polar region involved Brodmann's area 10, and the cingulate region appeared to involve the pregenual part of area 32. A fourth area of increased BF in the PFC was identified in the right frontal polar cortex. All of these regions are heavily interconnected with the ipsilateral amygdala (*1*).

The depressed group also had abnormally increased BF in the left medial thalamus and decreased BF in the medial caudate bilaterally (*13*). The data in the caudate were consistent with Baxter *et als.*' finding (*3*) of decreased metabolic rates for glucose (MRglu) in the caudate in unipolar depression (MRglu is presumably coupled with BF in the resting state). The result in the medial thalamus was replicated in our laboratory using MRglu measurements (see below).

The relationships between activity in these regions and depression severity were explored using correlations between regional BF and total HDRS scores. The HDRS scores correlated positively with amygdala BF ($r=0.73$, $p<.05$), but negatively with left prefrontal BF ($r=-0.62$, $p<.05$). No significant correlations were found between anxiety ratings and BF in the amygdala ($r=0.045$) or the left PFC ($r=0.32$).

1. Comparisons Between FPDD and non-FPDD Unipolar Depressives

The regional BF data in the depressed FPDD subjects were compared with those obtained using the same methods in a group of unipolar depressed subjects who did not meet criteria for FPDD, but were similar in mean age, gender composition and depression ratings ($n=8$; mean age$=28.1\pm6.0$ (range 21–36); 71% female; mean HDRS score$=29.9\pm8.0$ (range 23–43)). Mean normalized BF appeared increased in the FPDD group relative to the non-FPDD unipolar depressed group in the amygdala (4.7%, $t=2.6$, $p<.05$), the left PFC (4.9%; $t=1.8$, $p<0.10$) and the medial thalamus (5.8%; $t=2.0$, $p<0.10$). Differences between the non-FPDD group and the control group were not significant in any of these ROI (region of interest).

2. Replication Using PET Measurements of Glucose Metabolism

We replicated many of these findings in a new sample of primary unipolar depressives using MRglu measurements. Image data were acquired with a Siemens/CTI ECAT 953B/31 tomograph (in-plane resolution$=5$ mm FWHM, axial resolution$=5$ mm FWHM with septa inserted) and 10 mCi of ^{18}F-DG in depressed subjects with FPDD ($n=11$) or non-FPDD unipolar depression ($n=11$) and controls ($n=10$). The FPDD and non-FPDD subjects were similar in mean age (39 ± 7 and 44 ± 13, respectively). HDRS scores (25 ± 4 and 22 ± 4, respectively) and gender (73% and 64%, respectively). They significantly differed, however, in Automatic Thought Questionnaire (ATQ) scores (73 ± 22 and 51 ± 23, respectively; $p<0.05$). Subjects were unmedicated ≥4 weeks.

The metabolic differences between the FPDD and control groups were similar in magnitude and direction to the previously reported BF differences (*13*). Metabolism in the left PFC was increased in the FPDD group relative to both the control group ($t=2.2$, $p<0.05$) and the non-FPDD depressed group ($t=2.2$, $p<0.05$). The mean $\pm$ SD regional MRglu/global MRglu was 1.22 ± 0.102, 1.11 ± 0.142, and 1.16 ± 0.082 in the FPDD, non-FPDD and control groups, respectively. This difference between the two depressed groups could conceivably relate to the difference in negative ruminations as reflected by the ATQ scores (*13*). In the medial thalamus the difference between the FPDD and control groups was significant ($t = 4.3$, $p < 0.0005$) and the mean MRglu for the non-FPDD group did not significantly differ from either the control or the FPDD group (1.13 ± 0.097, 1.04 ± 0.181 and 0.954 ± 0.089 in the FPDD, non-FPDD, and control groups, respectively). In the amygdala, MRglu was significantly increased in the entire unipolar depressed sample relative to the control group (0.845 ± 0.074 and 0.779 ± 0.083, respectively, $p < 0.05$) and the difference between the FPDD and non-FPDD depressed groups was not significant.

3. *Correlations with Illness Severity*

Similar to the clinical correlations with regional BF (*13*), in FPDD the HDRS scores correlated positively with left amygdala metabolism ($r = 0.65$, $p = .04$) and negatively with left PFC metabolism ($r = -0.72$, $p = .06$). In addition, amygdala MRglu correlated negatively with left PFC metabolism ($r = -0.65$, $p = .03$). Moreover, the HDRS scores and the amygdala-to-PFC metabolic ratio were strongly correlated ($r = -0.79$, $p = 0.01$), which compared with $r = 0.81$ for our previous data in the FPDD subjects scanned with PETT VI. Finally, left prefrontal flow correlated negatively with depressive ideation as measured by ATQ scores ($r = -0.58$, $p = .08$). As in the PETT VI study correlations with anxiety (Spielberger State Anxiety Inventory scores) were not significant in the left PFC ($r = -0.20$) or the left amygdala ($r = 0.28$).

The direction of the correlations between depression ratings and regional activity in FPDD paralleled those found in the PET studies of emotion. During sadness induced in control subjects by a task that involved viewing sad faces, BF increased in the orbital cortex and the left amygdala, with sadness ratings correlating positively with amygdala BF and negatively with orbital BF (*49*). Similarly, during sadness induced by contemplation of sad thoughts or memories (*11*) we demonstrated that BF increased in parts of the left VLPFC and orbital coretex, with ΔBF negatively correlated with sadness ratings (notably, amygdala BF decreased during this task, consistent with other indications that, in normal subjects, amygdala BF increases are not observed with PET unless emotionally charged sensory stimuli are perceived during scanning (*11*). Moreover, in obsessive-compulsive subjects scanned during exposure to phobic stimuli, BF increased in the right posteromedial orbital (PMOrb) cortex, with ΔBF correlating negatively with Obsessive-Compulsive Analog scores ($r = -0.83$) (*43*). This finding is compatible with our observation in FPDD that ventral prefrontal metabolism is abnormally increased during the depressed state but correlates negatively with ATQ scores ($r = -0.58$). Finally, we recently found that in subjects with animal phobias scanned during repeated exposures to a phobic stimulus, BF increased in the left PMOrb during exposures that occurred after the development of habituation relative to BF measured during exposures that preceded habituation (mean ΔBF$=2.9\%$, $t=2.7$, $p<0.05$) or during rest (mean ΔBF$=3.3\% \pm 2.3\%$; $t=3.8$, $p<0.005$). Thus ventral prefrontal cortical areas become activated in a variety of emotional states, but prefrontal activity correlates inversely with ratings of negative thought or emotion.

4. Functional Correlations with Brain Activity

The relationships between regional BF and depression severity in FPDD appear consistent with evidence from single unit and lesion studies indicating that the amygdala is involved in forming stimulus-reinforcement associations while the ventral PFC participates in correcting behavioral responses that become inappropriate as reinforcement contingencies change. The amygdala plays a role in assigning affective significance to experiential stimuli (*28*, *39*) and the associations formed in the amygdala between reinforcement and sensory stimuli are long-lasting (*28*, *35*). Extinction of the responses to these associations may depend upon projections from the ventral PFC. For example, in the rat, lesions of the medial and orbital PFC prevented extinction to a conditioned stimulus (CS), but had no effect upon the rate of acquisition of fear conditioning (*35*).

The ventral PFC contains relatively high proportions of cells that undergo sustained change in their rate of discharge during the delay period between a reward-linked sensory cue and the subsequent response to it, with activity following this delay period being related to the presence or absence of reward (*17, 21, 45, 46, 58*). This system appears capable of rapid adaptation to changing reward contingencies, as the associations formed by orbital neurons between sensory stimuli and availability of reward show one trial stimulus-reinforcement association reversal (*45, 58*). Lesions of the ventral PFC thus result in perseveration of responses to nonreinforced stimuli. For example, monkeys with surgical lesions of the inferior convexity (the VLPFC and lateral orbital cortices) demonstrate "perseverative interference", characterized by difficulty in learning to withhold responses to nonrewarding stimuli (*23*) and humans with lesions of the PFC exhibit difficulty shifting intellectual strategies in response to changing demands and perseverate in strategies that become inappropriate (*17, 21*). Ventral prefrontal damage may also result in perseveration within the emotional domain, since medial prefrontal and orbital damage interferes with extinction of conditioned fear responses during nonreinforcement (*35*).

The apparent functions of the amygdala and the ventral PFC suggest a model of depression in which the amygdala assigns or projects associations between negative emotion and thoughts, social interactions or sensory stimuli, while the ventral PFC mediates attempts to break those associations and convert nonrewarding thoughts or behaviors to rewarding ones. This model could account for biological heterogeneity, since the depressive syndrome could arise from dysfunction within either the amygdala or the PFC. For example, dysfunction of the amygdala in FPDD may yield a state where negative affective labels are inappropriately assigned to all stimuli or where neurons storing negative associations are excessively or inappropriately activated (*19*), and the left ventral PFC is activated to break these nonrewarding associations. In contrast to FPDD, other depressive subtypes may be associated with dysfunction within the ventral PFC so that the ability to appropriately shift emotional or cognitive sets is impaired, resulting in a perseverative state involving melancholic thoughts or emotions. For example, cerebrovascular lesions of the left PFC increase the risk for developing major depression following stroke (*54*).

However, this hypothesis is overly simplified in regards to several issues. First, the finding of increased prefrontal BF and metabolism in FPDD does not indicate whether the functional output of that area is normal or abnormal. It is possible that in FPDD increased activity in the VLPFC or the orbital cortices is associated with abnormal modulation of those areas which interferes with the ability to shift emotional or cognitive sets, such that a negative thought pattern or mood is maintained. Second, the cytoarchitectonic subdivisions within the ventral PFC and the various nuclei of the amygdala likely have diverse and specialized functions that are not individually considered by this simplified model (*5*). Third, while BF was increased in the VLPFC and the orbital cortices in FPDD, metabolism was decreased in other prefrontal areas (Fig. 3B), highlighting the importance of approaching the ventral PFC as a functionally heterogenous region.

5. Anatomical Circuits Related to FPDD

In addition to considering the functional complexity within regions where metabolic abnormalities are observed, their connectivity must be considered. Our findings of increased BF in the amygdala and the ventral PFC coupled with indications of increased flow in the medial thalamus and reduced flow in the medial caudate support a variety of indirect evidence suggesting that limbic cortico-striatal-pallidal-thalamic circuits are involved in the pathophysiology of depression (*15*, *33*, *55*). Our data specifically suggested that two interconnected circuits are involved in the pathophysiology of FPDD: a limbic-thalamo-cortical circuit involving the amygdala, the mediodorsal nucleus (MD) of the thalamus (in the medial thalamus) and the ventrolateral and medial PFC; and a limbic-striatal-pallidal-thalamic circuit involving related parts of the striatum and the ventral pallidum as well as the components of the other circuit.

The first of these circuits can be conceptualized as an excitatory triangular circuit whereby the amygdala and the prefrontal cortical regions are interconnected by excitatory projections with each other and with MD (*1*, *27*, *47*). Increased metabolic activity in the amygdala, the medial thalamus and the PFC would presumably reflect increased synaptic activity in these areas, suggesting that synaptic transmission through the limbic-thalamo-cortical circuit is pathologically elevated in FPDD (*42*). Consistent with this hypothesis, neurosurgical procedures that ameliorate treatment-refractory depression interrupt projections within the limbic-thalamo-cortical and the limbic-striatal-pallidal-thalamic circuits (*2, 25, 36, 38*) and depressed patients referred for such treatments have increased concentrations of aspartic acid in the ventral frontal lobe (*16*). Other somatic AD treatments may also compensate for the pathophysiology of FPDD by suppressing limbic-thalamo-cortical activity, as we demonstrated that antidepressant drug treatment is associated with decreased BF and MRglu in the amygdala, the ventral PFC, and the medial thalamus (*10, 12*).

The implication of limbic cortical-striatal-pallidal-thalamic circuits in the pathophysiology of depression provides another framework from which to approach biological heterogeneity within mood disorders. Both the PET imaging differences between primary unipolar and bipolar depression (*11*) and the effects of neuropathological conditions that induce major depression suggest that distinct depressive subtypes may be associated with dissimilar functional anatomical abnormalities within these circuits (*13*). For example, lesions that involve the parts of the PFC that participate in the limbic-thalamo-cortical circuit and diseases of the basal ganglia (*e.g.* Parkinson's Disease and Huntington's Disease) are associated with higher rates of major depression than other similarly debilitating conditions (*15, 24, 31, 54*). These observations suggest that dysfunction at multiple points within this system may give rise to the major depressive syndrome. Because these conditions affect this system in different ways, it may also be suggested that imbalances within these circuits, rather than overall increased or decreased synaptic activity within a particular structure, may be associated with the major depressive syndrome (*11, 15, 33*). We pursued this hypothesis by comparing the patterns of BF abnormalities in the prefrontal, thalamic, and basal ganglia regions involved in the limbic cortical-striatal-pallidal-thalamic circuitry between bipolar depressives and unipolar depressives with FPDD.

II. THE FUNCTIONAL ANATOMY OF BIPOLAR DEPRESSION

Bipolar-depressed (BP-D) subjects who met DSM III-R criteria for bipolar disorder, depressed phase were selected. The HDRS scores were in the moderately-to-severely depressed range, similar to the comparator FPDD groups (mean score $=24.2 \pm 3.5$ (20–29) for subjects scanned with PETT VI and 21.5 ± 4.3 (16–27) for subjects scanned with the 953B). Subjects had not received psychotropic medications or other drugs likely to alter cerebral BF for >4 weeks. One subject had previously been medicated with antipsychotic agents and one with lithium. Subjects with a history of substance abuse or dependence were excluded.

BF was measured in the regions where abnormalities were found in FPDD (the left VLPFC, the left amygdala, the left medial thalamus and the left caudate) and compared across the BP-D ($n=11$), the FPDD ($n=13$) and the control groups ($n=24$) scanned with PETT VI using MANOVA. Significant overall category effects existed between groups (Wilks' lambda $=0.555$; $F=3.68$; Num $DF=8$; $p=0.0010$). Pairwise contrasts between individual groups demonstrated that mean flow was increased in the FPDD group relative to the BP-D group in the medial thalamus (7.9%; $p<0.001$), the amygdala (6.9%; $p<0.05$), and the VLPFC (4.9%; $p<0.05$) (Fig. 2). Mean flow did not differ between the BP-D and control groups in these regions. These data supported previous indications that metabolic abnormalities differ between unipolar and bipolar depressives (*4, 6, 51*).

In the BP-D group an abnormality was identified in the portion of the PFC ventral to the genu of the corpus callosum, an area where we had not found abnormal activity in the unipolar depression studies. Mean BF in the subgenual PFC (SGPFC) was decreased 6.6% ($p<0.001$) in the bipolar depressives relative to the controls. Moreover, both regional BF and MRglu were decreased in this ROI (−12.6% ($p<0.025$) and −11.8% ($p<0.01$), respectively; Fig. 3B) in an independent set of bipolar depressives ($n=6$) and controls ($n=10$) scanned with a Siemens/CTI 953B/31 (FWHM = 5 mm for the MRglu images and 14 mm for the BF images after filtering).

Six of the BP-D subjects (all scanned with PETT VI) were

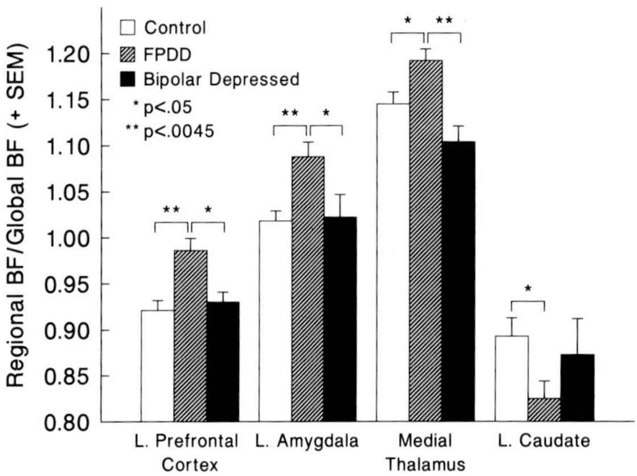

Fig. 2. Mean normalized BF for FPDD-depressed, bipolar-depressed, and control groups in the areas of the left PFC, amygdala, medial thalamus, and caudate where abnormalities had been demonstrated in FPDD (*13*).

subclassified as bipolar Type I, and eleven as Type II (by definition, Type I subjects have been hospitalized or incarcerated for mania (*22*)). Post hoc comparisons between BP-D Type I and II subjects showed that BF in the SGPFC was decreased in bipolar Type I subjects ($n=6$) relative to Type II subjects ($n=5$) for the image data acquired with PETT VI (mean normalized flow$=0.951\pm0.045$ and 1.01 ± 0.029, respectively; $p<0.05$).

Regional BF data in the SGPFC was also decreased in the BP-D group relative to the FPDD group (mean activity$=0.978\pm0.048$ and 1.04 ± 0.040, respectively; $p<0.05$; Fig. 3A). However, mean BF in the SGPFC was similar in the BP-D Type II and FPDD groups (1.04 ± 0.040 and 1.01 ± 0.029, respectively, for the PETT VI data and 1.06 ± 0.084 and 1.03 ± 0.093, respectively, for the 953B data) and in the FPDD subjects scanned with the 953B (which has higher spatial resolution than PETT VI) mean normalized MRglu was significantly decreased in the FPDD group ($n=11$) relative to the control group (1.06 ± 0.084 and 1.18 ± 0.125, respectively; $p<0.05$; Fig. 3B).

An additional abnormality was identified in the medial rostroventral thalamus, where BF was decreased in the BP-D group relative

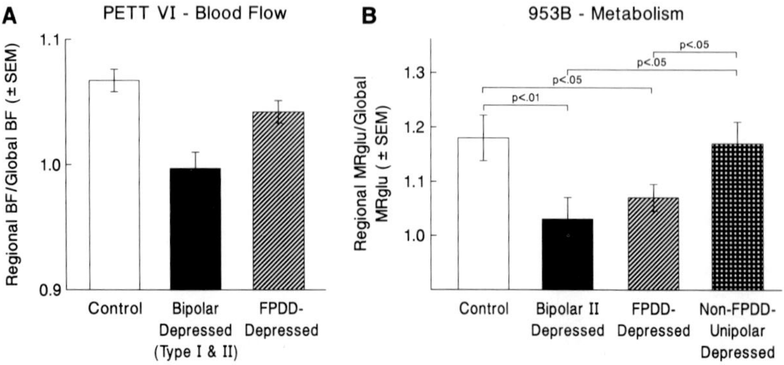

Fig. 3. A. Mean normalized BF was decreased in the SGPFC in the BP-D group relative to both the control group and the FPDD group. However, when the Type II subjects from the BP-D group were considered separately, their mean BF was similar to that of the FPDD group (see text). B. Mean normalized MRglu was also significantly decreased in the SGPFC in a separate sample of BP-D and control subjects scanned with a Siemens/CYI ECAT 953B/31 tomograph. MRglu was also decreased in the FPDD group relative to the control group and, as in the case of the PETT VI data, the difference between the BP-D Type II subjects and the FPDD subjects was not significant (all of the bipolar subjects scanned with the 953B were Type II). The 953B camera has higher resolution than PETT VI, so the magnitude of the differences appear greater than those in A.

to both the control group ($p<0.05$) and the FPDD group ($p<0.02$; mean flow was 1.08 ± 0.044, 1.13 ± 0.044, and 1.15 ± 0.063 in these groups, respectively). However, this ROI encompasses part of the third ventricle, which is reportedly enlarged in bipolar disorder (*8*, *57*). Since third ventricle enlargement could produce "partial volume effects" in the MRVTh in PET images (*32*), this morphometric abnormality could account for our finding of reduced flow in the MRVTh in BP-D. Moreover, since third ventricle enlargement was reported in bipolar but not unipolar depressives, this could account for the BF difference in the MRVTh in BP-D relative to FPDD.

Finally, an abnormality that was unique to bipolar Type I subjects was found in the lentiform nuclei, bilaterally. Considered alone, the Type I group had decreased flow relative to the Control group in both the right (mean decrease of -7.2%, $p<0.002$) and the left lentiform nuclei (mean decrease of -5.7%, $p<0.001$). Mean flow also differed between Type I and Type II subjects in the right lentiform nucleus (1.12 ± 0.042 and 1.24 ± 0.086, respectively; $p < 0.05$) and the left lentiform nucleus (1.15 ± 0.045 and 1.23 ± 0.083, respectively; $p < 0.05$). The most striking clinical difference between subgroups was the presence of delusions in all Type I but no Type II subjects. This may be relevant to the BF difference in the lentiform nucleus, since abnormal BF was reported in the pallidum in schizophrenia, where BF also correlated with psychotic features (*14, 30*).

Significance of the Subgenual Prefrontal Cortex in Depression

In architectonic terms the SGPFC consists of agranular cortex corresponding to Brodmann areas 32 and 25 in the prelimbic portion of the anterior cingulate gyrus (*5*). These areas have extensive connections with the amygdala, the lateral hypothalamus and the midbrain (*1, 5, 29, 52*). An abnormality within the SGPFC in bipolar depression could disrupt functional interactions between the SGPFC and these structures, potentially contributing to the emotional and neurovegetative manifestations of depression.

For example, the connections between the SGPFC and the hypothalamus could at least partially underlie the neurovegetative (*i.e.*, sleep, appetite, libido, and energy disturbances), neuroendocrine, and autonomic manifestations of depression (*7, 22, 37*). In addition, dysfunction of the connections between the SGPFC and the amygdala may result in perseveration of nonrewarded behavior and thought given the aforementioned evidence that lesions which included the SGPFC (although these large medial prefrontal lesions were not restricted to the SGPFC) prevented extinction of fear responses to a conditioned stimulus that was no longer reinforced (*35*). Finally, the midbrain connections include projections to dopaminergic neurons in the ventral tegmental area (A10) and the substantia nigra (A9) (*29, 52, 53*) which may participate in modulating DA release in the ventral striatum and the medial PFC (*18, 53, 56*). These projections are postulated to subserve a "reward-related system" that mediates motivation, hedonia, behavioral reinforcement, and psychomotor activity (*20, 40, 50*). Pharmacologic, neurochemical, and behavioral evidence suggests that the function of this system is diminished in both unipolar and bipolar depression, potentially reflected by anhedonia, amotivation, and psychomotor slowing (*22, 60*).

III. DISCRIMINATORY CAPABILITY OF REGIONAL BF MEASUREMENTS

To assess the sensitivity of PET measures of regional BF for classifying individual subjects into FPDD, BP-D, or control categories we performed a discriminant analysis of the BF data acquired with PETT VI in the PFC, caudate, thalamus, and amygdala. The FPDD ($n=13$), BP-D ($n=11$), and non-FPDD unipolar ($n=8$) groups used in this analysis did not significantly differ in mean age or depression ratings. In addition to the resting controls ($n=24$), controls ($n=7$) scanned during a sadness induction task involving the contemplation of sad thoughts (as a control for negative ruminations and sad mood during the scan) were also classified.

The discriminant analysis correctly classified FPDD subjects in 92% of cases and BP-D subjects in 73% of cases (100% of the bipolar Type I subjects were classified as BP-D). None of the BP-D subjects were misclassified as FPDD and none of the FPDD subjects were misclassified as BP-D. Both the resting and the "sad" controls were correctly classified as controls (in 88% and 84% of cases, respectively). The non-FPDD unipolar depressives were classified into the normal group (71%) or the BP-D group (29%).

Clinical Implications of Subtype

In addition to the family history and metabolic differences between the FPDD and non-FPDD unipolar groups, a treatment outcome distinction was identified. Of subjects who responded to selective serotonin reuptake inhibitor (SSRI) treatment at 8 weeks, the relapse rates during 6 months of uninterrupted treatment were 65% and 18% in the FPDD ($n=17$) and non-FPDD ($n=11$) groups, respectively ($p<0.05$ by Chi-squared). In contrast, only 1 of 8 FPDD subjects (13%) who responded to and continued treatment with tricyclic antidepressants relapsed during the same time period.

Besides the history of mania in the BP-D group, the bipolar and unipolar groups also differed in the risk for developing psychosis. All of the BP-D Type I subjects ($n=6$) were delusional at the time of scanning. In contrast, only one of the FPDD subjects and none of the BP-D Type II subjects scanned with either the PETT VI or the 953B cameras had ever been delusional. Pharmacologic differences between unipolar and bipolar depressives are described elsewhere (*22*).

SUMMARY

We investigated the functional anatomy of depression using PET measurements of cerebral BF and metabolism from unipolar depressives with FPDD, unipolar depressives who failed to meet criteria for FPDD, and bipolar depressives. In the depressed phase of FPDD, BF and metabolism were abnormally increased in the left PFC, the left amygdala and the left medial thalamus and decreased in the caudate. The affected area within the left PFC included the VLPFC, the frontal polar cortex, the medial orbital cortex and the pregenual portion of the anterior cingulate gyrus. In contrast to the BF and metabolic increases in these prefrontal areas, metabolism was decreased in the SGPFC (in the prelimbic portion of the anterior cingulate gyrus), an area where flow and metabolism were also decreased in BP-D. In the left VLPFC, BF was elevated in the depressed but not the unmedicated-remitted phase, while in the left amygdala flow appeared increased in both phases. Effective antidepressant drug treatment was associated with decreased BF and metabolism in VLPFC, amygdala, and medial thalamus.

In bipolar depression BF was decreased in the SGPFC, the MRVTh and, in Type I patients, the lentiform nucleus. Mean BF was not abnormal in the areas of the VLPFC, caudate, thalamus, or amygdala where abnormalities were identified in the unipolar depressives. These data appeared compatible with evidence from lesion studies indicating that the abnormalities within limbic cortical-striatal-pallidal-thalamic circuits that result in secondary bipolar disorders are distinct from those associated with secondary unipolar depressive disorders. They are also consistent with genetic, neurochemical, and pharmacologic evidence that unipolar and bipolar depression are pathophysiologically distinct.

REFERENCES

1 Amaral, D.G. and Price, J.L. (1984). *J. Comp. Neurol.* **230**, 465–496.
2 Ballantine, H.T., Bouckoms, A.J., Thomas, E.K., and Giriunas, I.E. (1987). *Biol. Psychiatry* **22**, 807–819.
3 Baxter, L.R., Phelps, M.E., and Mazziotta, J.C. (1985). *Arch. Gen. Psychiatry* **42**, 441–447.
4 Buchsbaum, M.S., Wu, J., DeLisi, L.E., Holcomb, H., Kessler, R., Johnson, J., King, A.C., Hazlett, E., Langston, K., and Post, R.M. (1986). *J. Affective Dis.* **10**, 137–152.
5 Carmichael, S.T. and Price, J.L. (1994). *J. Comp. Neurol.* **346**, 366–402.
6 Cohen, R.M., Semple, W.E., Gross, M.D., Nordahl, T.E., King, A.C., Pickar, D., and Post, R.M. (1989). *Neuropsychopharmacology* **2**, 241–254.
7 Damasio, A.R., Tranel, D., and Damasio, H. (1990). *Behav. Brain Res.* **41**, 81–94.
8 Dewan, M.J., Haldipur, V., and Lane, E.E. (1988). *Acta Psychiatr. Scand.* **77**, 670–676.
9 Drevets, W.C. (1995). In *Personality and Psychopathology*, ed. Cloninger, C.R., Washington, D.C.: American Psychiatric Press, Inc., in press.
10 Drevets, W.C. and Raichle, M.E. (1992). *Psychopharmacol. Bull.* **28**, 261–274.
11 Drevets, W.C. and Raichle, M.E. (1995). In *The Cognitive Neurosciences*, ed. Gazzaniga, M., pp. 1153–1164. Cambridge, MA: MIT Press.
12 Drevets, W.C., Videen, T.O., MacLeod, A.K., and Raichle, M.E. (1993). *Soc. Neurosci. Abstr.*, 7.
13 Drevets, W.C., Videen, T.O., Price, J.L., Preskorn, S.H., Carmichael, S.T., and Raichle, M.E. (1992). *J. Neurosci.* **12**, 3628–3641.
14 Early, T.S., Reiman, E.M., Raichle, M.E., and Spitznagel, E.L. (1987). *Proc. Natl. Acad. Sci. U.S.A.* **84**, 561–563.
15 Folstein, S.E., Peyser, C.E., Starkstein, S.E., and Folstein, M.F. (1991). In *Psychopathology and the Brain*, ed. Carrol, B.J. and Barrett, J.E., pp. 65–75. New York: Raven Press.
16 Francis, P.T., Poynton, A., and Lowe, S.L. (1989). *Brain Res.* **494**, 314–324.
17 Fuster, J.M. (1989). In *The Prefrontal Cortex—Anatomy, Physiology, and Neuropsychology of the Frontal Lobe*, New York: Raven Press.
18 Gariano, R.F. and Groves, P.M. (1988). *N.Y. Acad. Sci.* 505–507.
19 Gloor, P., Olivier, A., Quesney, L.F., Andermann, F., and Horowitz, S. (1982). *Ann. Neurol.* **12**, 129–144.
20 Goeders, N.E. and Smith, J.E. (1983). *Science* **221**, 773–775.
21 Goldman-Rakic, P.S. (1989). In *Handbook of Physiology—The Nervous System*, ed. Mills, J. and Mountcastle, V.B., vol. V, pp. 373–417. Baltimore: Williams & Wilkins.
22 Goodwin, D.W. and Jamison, K.R. (1990). *Manic-depressive Illness*, New York: Oxford University Press.

23 Iverson, S.D. and Mishkin, M. (1970). *Exp. Brain Res.* **11**, 376–386.

24 Jeste, D.V., Lohr, J.B., and Goodwin, F.K. (1988). *Br. J. Psychiatry* **153**, 444–459.

25 Knight, G. (1972). In *Psychosurgery*, ed. Hitchcock, E., Laitinen, L., Vaernet, K., pp. 267–277. Springfield: Thomas.

26 Kupfer, D.J., Frank, E., Perel, J.M., Cornes, C., Mallinger, A.G., Thase, M.E., McEachran, A.B., and Grochocinski, V.J. (1992). *Arch. Gen. Psychiatry* **49**, 769–773.

27 Kuroda, M. and Price, J.L. (1991). *J. Comp. Neurol.* **303**, 513.

28 LeDoux, J.E. (1987). In *Handbook of Physiology—The Nervous System*, ed. Mills, J., Mountcastle, V.B., Plum, F., and Geiger, S.R., pp. 373–417. Baltimore: Williams & Wilkins.

29 Leichnetz, G.R. and Astruc, J. (1976). *Brain Res.* **109**, 455–472.

30 Liddle, P.F., Friston, K.J., Frith, C.D., Hirsch, S.R., Jones, T., and Frackowiak, R.S.J. (1992). *Br. J. Psychiatry* **160**, 179–186.

31 Mayeux, R. (1983). In *Neuropsychology of Human Emotion*, ed. Heilman, K.M. and Satz, P., pp. 141–164. New York: Guilford.

32 Mazziotta, J.C., Phelps, M.E., Plummer, D., and Kuhl, D.E. (1981). *J. Comput. Assist. Tomogr.* **5**, 734–743.

33 McHugh, P.R. (1989). *Neuropsychiatr. Neuropsychol. Behav. Neurol.* **2**, 239–247.

34 Moldin, S.W., Reich, T., and Rice, J.P. (1991). *Behav. Genet.* **21**, 211–242.

35 Morgan, M.A., Romanski, L.M., and LeDoux, J.E. (1993). *Neurosci. Lett.* **163**, 109–113.

36 Nauta, W.J.H. (1973). In *Surgical Approaches in Psychiatry*, ed. Laitinen, L.V. and Livington, K.E., pp. 303–314. Baltimore: University Park Press.

37 Neafsey, E.J., Terreberry, R.R., Hurley, K.M., Ruit, K.G., and Frysztak, R.J. (1993). In *Neurobiology of Cingulate Cortex and Limbic Thalamus*, ed. Vogt, B.A. and Gabriel, M., pp. 206–233. Boston: Birkhauser.

38 Newcombe, R. (1982). In *Stereotaxy of the Human Brain*, ed. Schaltenbrand, G. and Walker, A.E., pp. 591–599. New York: Thieme-Stratton.

39 Nishijo, H., Ono, T., and Nishino, H. (1988). *J. Neurosci.* **8**, 3570–3583.

40 Phillips, A.G. and Fibiger, H.C. (1978) *Canad. J. Psychol.* **32**, 58–66.

41 Price, J.L., Carmichael, S.T., and Drevets, W.C. In *Progress in Brain Research*, in press.

42 Raichle, M.E. (1987). *Handbook of Physiology—The Nervous System*, vol. V, Baltimore: Williams & Wilkins.

43 Rauch, S.L., Jenike, M.A., Alpert, N.M., Baer, L., Breiter, H., Savage, C.R., and Fischman, A.J. (1994). *Arch. Gen. Psychiatry* **51**, 62–70.

44 Rice, J., Reich, R., Andreasen, N., Endicott, J., VanEerdewegh, M., Fishman, R., Herschfelt, R., and Klearman, G. (1987). *Arch. Gen. Psychiatry* **44**, 441–447.

45 Rolls, E.T. (1995). In *The Cognitive Neurosciences*, ed. Gazzaniga, M.S., pp. 1091–1106. Cambridge, MA: MIT Press.

46 Rosenkilde, C.E., Bauer, R.H., and Fuster, J.M. (1981). *Brain Res.* **209**, 375–394.

47 Russchen, F.T., Amaral, D.G., and Price, J.L. (1987). *J. Comp. Neurol.* **256**, 175–210.
48 Schatzberg, A.F., Samson, J.A., Bloomingdale, K.L., Orsulak, P.J., Gerson, B., Kizuka, P., Cole, J., and Schildkraut, J. (1989). *Arch. Gen. Psychiatry* **46**, 260–268.
49 Schneider, F., Gur, R.C., Gur, R.E., Harper-Mozley, L., Smith, R., Mozley, P.D., and Alavi, A. (1994). APA Annu. Meet.
50 Schultz, W., Apicella, P., Scarnati, E., and Ljungberg, T. (1992). *J. Neurosci.* **12**, 4595–4619.
51 Schwartz, J.M., Baxter, L.R., Mazziota, J.C., Gerner, R.H., and Phelps, M.E. (1987). *JAMA* **258**, 1368–1373.
52 Sesack, S.R., Deutch, A.Y., Roth, R.H., and Bunney, B.S. (1989). *J. Comp. Neurol.* **290**, 213–242.
53 Sesack, S.R. and Pickel, V.M. (1992). *J. Comp. Neurol.* **320**, 145–160.
54 Starkstein, S.E. and Robinson, R.B. (1989). *Br. J. Psychiatry* **154**, 170–182.
55 Swerdlow, N.R. and Koob, G.F. (1987). *Behav. Brain Sci.* **10**, 197–245.
56 Taber, M.T. and Fibiger, H.C. (1993). *Neuropsychopharmacology* **9**, 271–275.
57 Tanaka, Y., Hazama, H., and Fukuhara, T. (1982). *Folia Psychiatr. Neurol.* **36**, 137–144.
58 Thorpe, S.J., Rolls, E.T., and Maddison, S. (1983). *Exp. Brain Res.* **49**, 93–115.
59 Wang, R.Y. and Aghajanian, G.K. (1980). *Commun. Psychopharmacol.* **4**, 83–90.
60 Willner, P. (1995). In *Psychopharmacology: The Fourth Generation of Progress*, ed. Bloom, Kupfer, D.L., pp. 921–931. New York: Raven Press.
61 Winokur, G. (1982). *Pharmacopsychiatry* **16**, 620–633.

MEMORY

5

REGIONAL VARIATIONS IN THE HIPPOCAMPAL PLACE CODE

MIN W. JUNG

Department of Psychology and Arizona Research Laboratories, Division of Neural Systems, Memory and Aging, University of Arizona, Tucson, AZ 85724, U.S.A.

Following the seminal report of the case of patient H.M., who suffered from severe anterograde amnesia following bilateral removal of the hippocampus and other temporal lobe structures (*23*), there have been a number of studies on the possible role of the hippocampus in learning and memory. Although the exact nature of hippocampal function is still a matter of controversy, it is generally agreed that the hippocampus plays an important role in spatial memory in humans and other animals (*17*). The relationship between the hippocampus and spatial memory has been particularly well studied in the rat. The rat has a simpler nervous system than primates, and much more is known about the physiology of the rat hippocampus than about the primate hippocampus. Thus, investigations using the rat model may lead to a better understanding of the neural basis of spatial memory, and possibly other memory functions subserved by the hippocampus, in humans and primates.

One key discovery in this field is the existence of "place cells" in the rat hippocampus (*16*). Principal cells of the hippocampus are highly active in small areas of a given environment and very quiet outside these areas. Although influenced by non-spatial factors, place-specific firing is by far the strongest, most reliable known behavioral correlate of the principal cells in the rat hippocampus. Place-specific firing cannnot be linked to the animal's ongoing behavior or immediate sensory inputs in any obvious way, but the firing field of a place cell usually follows the rotation of distal visual cues, suggesting that information about allocentric spatial location with respect to distal landmarks is encoded by hippocampal neurons (*17*). Nevertheless, place-specific firing is maintained following deprivation of one or more sensory modalities, including vision (*6*), indicating that multiple sensory inputs contribute to place-specific firing. Results from these experiments suggest that place cells reflect an allocentric place code that is constructed from highly-processed, multiple sensory inputs, although this is by no means a clearly resolved issue.

Previous studies of place cells have contributed greatly to our knowledge about the nature of spatial information represented in the hippocampus. However, these recordings were performed mostly in dorsal CA1 or CA3, and relatively little attention has been paid to regional differences in place cell properties, leaving spatial firing patterns of the units in the dentate gyrus (DG) and ventral hippocampus largely unknown. The DG is a major structure of the hippocampus, and damage to this area profoundly impairs spatial memory (*25*). Thus information about the behavioral correlates, especially place-specific firing, of units in this area is important for understanding the neural basis of hippocampal spatial memory. Regarding the ventral hippocampus, one might assume that the functional role and operational principles are similar to those of the dorsal hippocampus, since the gross anatomy is similar. Various studies have shown, however, wide variations in anatomy, neuropharmacology and physiology along the septotemporal axis, suggesting functional differences between the dorsal and ventral hippocampus. It is therefore of interest to compare behavioral correlates of ventral hippocampal units with those of dorsal hippocampal units.

In this chapter, recent experimental results will be summarized that examined the spatial firing characteristics of dentate granule cells and ventral CA1 neurons. Also, characteristics of place-specific firing will be compared along the transverse and longitudinal axes of the hippocampus, and their theoretical implications will be discussed. For convenience, "hippocampus" is defined to include the DG, CA3, and CA1, but to exclude the subiculum in this chapter.

I. ARE GRANULE CELLS THETA CELLS?

Initial recordings in the DG suggested that granule cells, unlike CA1 or CA3 pyramidal cells, discharge at high baseline rates, have the characteristics of "theta" cells (inhibitory interneurons found throughout the hippocampus) (*20*), and have little spatial selectivity (*4, 22*). Pyramidal cells and inhibitory interneurons in CA3 and CA1 can be easily distinguished based on their spike discharge characteristics. Pyramidal cells fire in bursts and at low overall rates. Virtually all place cells are pyramidal cells. Inhibitory interneurons increase their discharge rates and fire in synchrony with theta (6–8 Hz) electroencephalogram (EEG) during locomotion, thus named theta cells. Their overall firing rate is much higher than that of pyramidal cells. The initial reports were therefore puzzling in that granule cells, which outnumber neurons in all other hippocampal regions combined, apparently behaved like inhibitory interneurons in Ammon's horn. Also, known physiological properties of the hippocampus do not fit snugly with this conclusion, as discussed in detail elsewhere (*13*). Furthermore, classification of units based on their correlation to perforant path-evoked population spike suggests that high rate units in the DG correspond to inhibitory interneurons, with granule cells firing at much lower rates (*15*). These considerations suggest the possibility that previous recordings may have missed most granule cells because they are largely silent, and that previously claimed granule cells may actually be inhibitory interneurons.

Rigorous reexamination of this issue strongly suggests that granule cells indeed fire at very low rates, and theta cells in the DG correspond to inhibitory interneurons in this region (*7*). In this study the "stereotrode" recording technique was used, which allows the on-line visualization of spike data history as distinct clusters, therefore facilitating the detection of low rate units. The most compelling evidence is the finding that most units do not show characteristics of theta cells when multiple units are simultaneously recorded. The majority of cells in the dentate granular layer are granule cells, whereas inhibitory interneurons comprise only a small subset. Thus, if granule cells are theta cells, the majority of simultaneously recorded units should exhibit characteristics of theta cells, including high baseline firing rates. However, most recorded units fired at low rates (well below 1 Hz) and in bursts. When the finding was replicated using a "tetrode" (*27*), which has four probes instead of two and is therefore more sensitive, as many as 15 units could be simultaneously recorded in the dentate granular layer. For each recording session, usually one or no theta cells were recorded. The majority were low rate, bursty units. There has been only one case in which two theta cells were simultaneously recorded (Jung and McNaughton, unpublished observation). Furthermore, low rate, bursty units were activated above the population spike threshold and inside the population spike time window in response to perforant path stimulation, whereas theta cells were usually activated outside the population spike time window and even in the absence of the population spike (when stimulation is weak), which is the typical activation pattern for inhibitory neurons in CA1 (*5*). The dentate population spike presumably reflects synchronous discharge of granule cells; therefore granule cells should be correlated with the population spike when stimulated. These and other experimental results (*7*) strongly favor the conclusion that theta cells in the dentate granular layer are inhibitory interneurons, and that real granule cells fire in bursts and at extremely low rates. In the following, low rate units recorded in the DG will be referred to granule cells, even though this identification is not unequivocally resolved.

II. SPATIAL FIRING CHARACTERISTICS OF DENTATE GRANULE CELLS

The granule cells clearly showed place-specific firing on an 8-arm radial maze. The spatial firing characteristics of the granule cells were essentially the same as CA3 place cells recorded in the same study, including spatial information content (*7*). Granule cells may therefore be called place cells in the same sense as CA1 or CA3 place cells. Like CA1 and CA3 pyramidal cells, not all granule cells were active on the maze. A preliminary estimate using tetrode recording suggests that less than 30% are active on an 8-arm radial maze (Jung and McNaughton, unpublished observation). One outstanding difference was that granule cells tend to have multiple, tight place fields whereas CA1 and CA3 place cells have single place fields in most cases. Some of the examples for granule cells are shown in Fig. 1A. It is noteworthy that Rose *et al.* (*22*) recorded 7 low rate units (out of 56), and 3 of them showed place-specific firing. These units fired in bursts and at substantially lower rates than theta cells, which is consistent with the result of Jung and McNaughton (*7*).

Theta cells in the DG, presumably inhibitory interneurons, fired at high rates all over the maze, with little spatial selectivity, which is

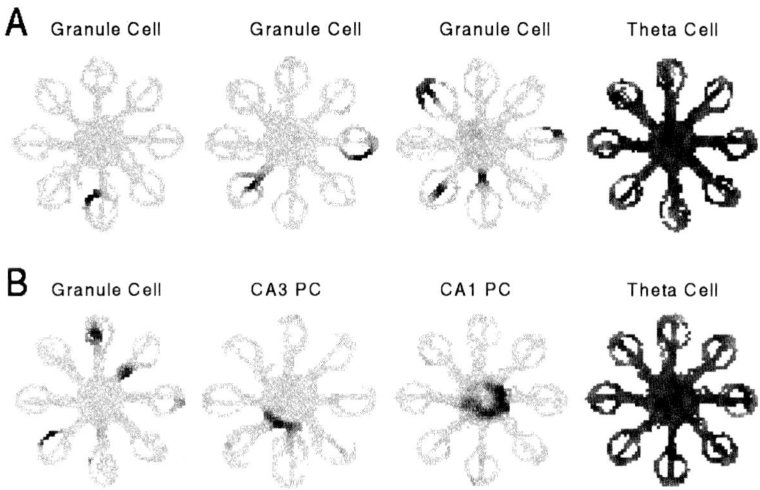

Fig. 1. Each plot is a top-down view of a spatial firing map for one cell on an 8-arm radial maze. Rats ran 10–12 trials of a spatial working memory task in 20–30 min. The cumulative trajectory of a rat is depicted by light gray and firing rate is indicated in gray scale, so that dark gray indicates high firing rate at that location. (A) Spatial firing maps for 3 granule cells (low rate units) and a theta cell (high rate unit) recorded in the dentate granular layer are shown. As shown, granule cells have one or more specific place fields, whereas theta cells fired at high rates all over the maze. (B) Representative examples of spatial firing map for each cell type in the dorsal hippocampus. As shown, granule cells have tight place fields with a tendency to have multiple fields; CA3 pyramidal cells (PC) usually have single, selective fields; CA1 pyramidal cells generally have single, fairly selective (but less selective than CA3 pyramidal or granule cells) fields; and theta cells in all fields fire at high rates all over the maze.

similar to the properties of CA1 theta cells. An example is shown in Fig. 1A.

III. COMPARISON OF PLACE CODE ALONG THE TRANSVERSE AXIS

The following general feature emerges for spatial firing patterns of units along the transverse axis of the hippocampus in a small, experimental environment (*1*, *7*, Markus *et al.*, unpublished observation) (see Fig. 1B): granule cells have highly selective place fields with a tendency for multiple fields; CA3 pyramidal cells mostly have single, highly selective fields; CA1 pyramidal cells usually have single, fairly selective fields, but less selective than CA3 pyramidal or dentate granule cells. Instead, in-field firing is more reliable for CA1 neurons. A caveat for this comparison is that there could be considerable variations for each cell type. Inhibitory interneurons throughout all fields fire at high rates all over the maze with slight spatial biases. This feature indicates both similarities and differences. The operational modes are similar in that principal cells (granule cells and pyramidal cells) fire at low overall rates with subsets of them showing selective place-specific firing, and inhibitory interneurons fire at high rates with low spatial selectivity. Also, principal cells are more active when an animal is in inactive states, whereas inhibitory interneurons increase their activities during locomotion, when the hippocampus exhibits strong theta frequency (6–8 Hz) EEG.

While the basic operational principles seem to be the same across the transverse axis, some differences are found in spatial firing patterns. First, granule cells tend to have multiple place fields whereas pyramidal cells mostly have single place fields. Second, the overall activity of granule cells is very low. Third, place-specific firing is less selective, but more reliable for CA1 than for CA3 or dentate place cells.

IV. THEORETICAL IMPLICATIONS

The above comparison indicates that highly specific spatial information is processed in the DG and CA3, and that specificity gets somewhat lower in CA1. This does not support the possibility that the spatial resolution of place-specific firing becomes progressively refined as one moves from the DG through CA3 to CA1 (an assembly line analogy). It does not fully support the "two-stage" model of memory trace formation (*3*) in its original form either. This model proposes that neocortical information is transmitted to the DG during exploratory behaviors (theta state), since granule cells in the DG were known as theta cells at that time, which induces weak and transient potentiation in CA3. During inactivity ("sharp wave" state), the weakly potentiated CA3 pyramidal neurons initiate sharp wave bursts that result in stable long-term potentiation (LTP) in CA3 and CA1. As described above, a more recent study (*7*) strongly suggests that the behavioral dependence of granule cells is similar to that of pyramidal cells, suggesting that all three subfields operate in the same mode during the theta as well as sharp wave state. This makes it unlikely that the DG is primarily involved in information processing during the theta state and then shuts off during the sharp wave state. Considering that theta frequency is nearly optimal to induce LTP, it seems more likely that information is initially encoded in the hippocampus in the form of LTP throughout the fields during the theta state, and memories are replayed and consolidated throughout the fields during the sharp wave state.

The above spatial firing patterns are consistent with some theoretical considerations. They indicate that spatial information is coded in a sparse manner throughout the hippocampus (*10*). Sparse coding means that small subsets of neurons are used to represent input patterns. This has effects of minimizing the usage of synapses and reducing probability of overlap among inputs, therefore increasing the maximal memory capacity of a neural network (*10*, *26*). Only subsets of pyramidal or granule cells are active during exploration, and units discharge briefly in their place fields and are almost silent outside. Therefore only a small proportion (on the order of 1%) of principal cells are active at a given moment, indicating that spatial information is sparsely coded in each field of the hippocampus. Unit recording studies indicate that the overall firing rate of granule cells is particularly low (well below 1 Hz) whereas neurons in the entorhinal cortex (EC) fire at much higher rates (well above 1 Hz) (*1*, *7*, *19*). Also, spatial selectivity of unit firing is much lower for the EC than the hippocampus, suggesting that the EC-DG projection accomplishes a high degree of sparsification. One factor for this operation may be the high discharge threshold of granule cells (*12*). These results are consistent with the proposal that the DG performs orthogonalization of input patterns (*11, 21*). By separating similar input patterns into distinct ones (increasing Hamming distance), more patterns can be stored and retrieved without error. It was proposed that similar patterns in the EC would be separated in the DG since there are more neurons (about 1 million) in the DG than in the EC (about 100,000) (*11*). Assuming that the same number of units is used to represent one pattern, overlap between two patterns is less likely in the DG (since there are more units) than in the EC. High sparsity in the DG would facilitate the proposed pattern separation.

The overall firing rate of principal cells during exploration is higher in CA1 than CA3 or DG (*1*, *7*, Markus *et al.*, unpublished observation), and a preliminary estimate indicates a larger proportion of active neurons in CA1 than CA3 or DG during exploration (Jung *et al.*, unpublished observation). These results imply that the number of active neurons at a given moment is higher for CA1 than the other regions, and therefore spatial information is coded in a less sparse manner in CA1. Conversely, reliability of unit firing in the place field (across multiple visits) is higher for CA1 than CA3 neurons (Markus *et al.*, unpublished observation). The functional significance of these differences is not clear at present. The differences may be an indication that CA1 participates in recoding of CA3 patterns into a less sparse, more reliable form before it projects to the neocortex. The anatomical structure of CA3 (massive recurrent collaterals) is consistent with theoretical requirements for storage of autoassociative memory (*10, 14, 21*), whereas CA1, which lacks recurrent collaterals and is located at the final stage of the trisynaptic circuit, is more likely to be an output device. The DG has been suggested to be involved in preprocessing of neocortical inputs (such as orthogonalization, see above) (*10, 11, 21*), or storage of associative memory itself (*2*). It is therefore possible that information is sparsified (redundancy is reduced) as one advances from the sensory cortex toward higher order association cortices, culminating in the DG-CA3 where patterns are associated and stored, and then redundancy increases (sparsity decreases but reliability increases) again as one advances to CA1, subiculum, EC and back to the neocortex. The experimental results are at least consistent with this possibility.

V. COMPARISON OF PLACE CODE ALONG THE LONGITUDINAL AXIS

Most theoretical considerations of hippocampal information processing concerned information flow along the transverse axis, probably due to the fact that the major subdivisions (DG, CA3, CA1) are found at any level along the longitudinal axis. Also, most single unit recordings have been performed at the dorsal (septal) hippocampus for the rat, mostly due to technical reasons, leaving the behavioral physiology of the ventral (septal) hippocampus largely unknown. Anatomical studies have shown, however, that there are wide variations in anatomical connectivity along the longitudinal axis of the hippocampus (*e.g.*, *28*). It is a premise of modern neuroscience that variations in neural connectivity are accompanied by variations in neural functions. Physiological and behavioral studies also indicate variations along the longitudinal axis. Jung *et al.* (*8*) recently com-

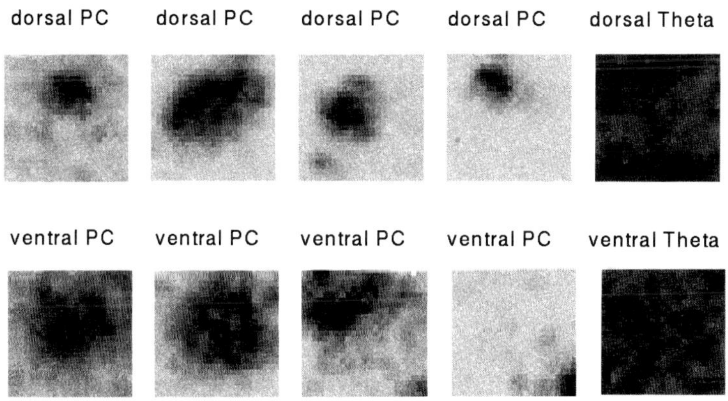

Fig. 2. Spatial firing maps for some dorsal and ventral pyramidal cells (PC) and theta cells (inhibitory interneurons). Firing rate is indicated in gray scale as in Fig. 1. As shown, dorsal pyramidal cells had fairly selective place fields, whereas ventral pyramidal cells showed a wider range of spatial selectivity, so that overall information content is lower for ventral pyramidal cells. Theta cells in both dorsal and ventral CA1 fired at high rates all over the maze.

pared spatial firing patterns of dorsal and ventral CA1 units while rats were performing a random food search and retrieval task in a square box (68 × 68 cm) containing a simple visual cue. Three conclusions were drawn from this study: 1) ventral units, like dorsal units, can be classified into complex spike and theta cells, and place cells exist in the ventral CA1; 2) the proportion of active units in the arena is much lower for ventral units than dorsal units; 3) the spatial resolution of the ventral place cells is significantly lower than that of the dorsal place cells. Some examples of dorsal and ventral place cells are shown in Fig. 2.

The third conclusion is in contrast to a similar study by Poucet *et al.* (*18*) which reports similar information content for dorsal and ventral CA1 units. There are several possible sources for the different conclusions. One is differences in experimental procedures. Poucet *et al.* selected recordable units in the arena, whereas Jung *et al.* selected units while rats were sleeping outside the arena not knowing *a priori* if the selected units would be place cells. Also, small variations in the recording sites may have resulted in the different results. More systematic studies will be required to resolve the discrepancy.

VI. FUNCTIONAL IMPLICATIONS

The result of Jung *et al.* (*8*) supports the emerging principle that there exist major heterogeneities along the longitudinal axis of the hippocampus. A number of possible functional interpretations can be suggested from the results. For example, the external space may be represented at different scales along the longitudinal axis, which could be computationally advantageous. The dorsal and ventral hippocampus may deal with high resolution spatial information for a small environment and low resolution information for a large environment, respectively. Alternatively, non-spatial information processing may be a preeminent role of the ventral hippocampus. On the basis of anatomical connectivity, Witter (*28*) has proposed that the dorsal hippocampus is preferentially involved in the processing of sensory related information, whereas the ventral hippocampus is involved in the processing of visceral related information. In this vein, it would be important to compare behavioral correlates of the dorsal *versus* ventral hippocampal units in various behavioral contexts, such as emotional learning.

VII. PLACE CELLS IN THE NEOCORTEX?

So far place cells have been found in the hippocampus (DG, CA3, and CA1), subiculum and EC, but not in the isocortex. Further, spatial selectivity is much lower for the units in the subiculum and EC (*1, 19, 24*), which are the major input/output structures for the hippocampus, suggesting that specific encoding of allocentric space is a special feature of the hippocampal formation. Some parts of the prefrontal cortex receive monosynaptic projections from the temporal portion of CA1 both in rat and monkey, and brief tetanization of this pathway induces LTP (*9*), suggesting that allocentric spatial information may be incorporated into the prefrontal cortex and may be exhibited in the form of place-specific firing. The author therefore recorded units in the prefrontal cortex while rats were performing an 8-arm radial maze, spatial working memory task. Among about 200 units recorded in the regions of the prefrontal cortex, including targets of hippocampal afferents, no place-specific firing was observed except when spatial locations were related to distinct phases of the tasks (Jung *et al.*, unpublished observation). This result supports the conclusion that place-specific firing is a distinct property of the hippocampal formation.

VIII. CONCLUSION

Currently there exists no theory that is generally regarded to capture the essential nature of hippocampal computation, with the details of hippocampal anatomy and physiology successfully simplified and incorporated. Optimistically, however, progress is being made at various levels of analysis, from biophysics of ion channels to computational theory and neuropsychology, which closely interact with each other and co-evolve. This chapter reviewed recent findings in behavioral physiology. We now know the basic spatial firing characteristics in each subfield of the dorsal hippocampus and are beginning to understand those of the ventral hippocampus. Obviously more information is required on discharge characteristics of ventral hippocampal neurons, including those of ventral CA3 and DG. Also, preliminary findings in the dorsal hippocampus, that were mentioned throughout this chapter, must be confirmed and quantified more rigorously. These data will serve as important constraints for the future endeavor of formulating computational theories of hippocampal information processing.

SUMMARY

Relatively little information has, until recently, been available about spatial firing patterns of dentate granule cells and ventral hippocampal neurons of the rat. Contrary to previous reports, recent results indicate that theta cells in the dentate granular layer are likely to be inhibitory interneurons and that granule cells exhibit highly selective place-specific firing. Thus, the basic operational principles appear to be the same across the transverse axis of the dorsal hippocampus (dentate gurus, CA3 and CA1), but some differences are found in their spatial firing patterns. These results provide new constraints on computational theories of the hippocampus.

Recent experiments demonstrated existence of place cells in the ventral CA1 area. However, a smaller proportion of ventral units were active than dorsal units during exploration, and overall spatial information was lower for ventral units. These results complement the emerging principle that there exist major heterogeneities along the longitudinal axis of the hippocampus. Ventral CA1 sends monosynaptic projections to the prefrontal cortex, suggesting possible existence of place cells in this region. However, no place cells were found among prefrontal cortical neurons. This result supports the notion that place-specific firing is a distinct property of the hippocampal formation.

REFERENCES

1 Barnes, C.A., McNaughton, B.L., Mizumori, S.J.Y., Leonard, B.W., and Lin, L.-H. (1990). *Prog. Brain Res.* **83**, 287–300.

2 Buckmaster, P.S. and Schwartzkroin, P.A. (1994). *Hippocampus* **4**, 393–402.

3 Buzsâki, G. (1989). *Neuroscience* **31**, 551–570.
4 Buzsâki, G., Leung, L.-W.S., and Vanderwolf, C.H. (1983). *Brain Res. Rev.* **6**, 139–171.
5 Fox, S.E. and Ranck, J.B., Jr. (1981). *Exp. Brain Res.* **41**, 399–410.
6 Hill, A.J. and Best, P.J. (1981). *Exp. Neurol.* **74**, 204–217.
7 Jung, M.W. and McNaughton, B.L. (1993). *Hippocampus* **3**, 165–182.
8 Jung, M.W., Wiener, S.I., and McNaughton, B.L. (1994). *J. Neurosci.* **14**, 7347–7356.
9 Laroche, S., Jay, T.M., and Thierry, A.M. (1990). *Neurosci. Lett.* **114**, 184–190.
10 Marr, D. (1971). *Philos. Trans. R. Soc. Lond. (Biol.)* **262**, 23–81.
11 McNaughton, B.L. (1989). In *Neural Connections, Mental Computations*, ed. Nadel, L., Cooper, L.A., Culicover, P., and Harnish, R.M., pp. 285–350. Cambridge, MA: MIT Press.
12 McNaughton, B.L., Barnes, C.A., and Anderson, P. (1981). *J. Neurophysiol.* **46**, 952–966.
13 McNaughton, B.L., Barnes, C.A., Mizumori, S.J.Y., Green, E.J., and Sharp, P.E. (1991). In *Kindling and Synaptic Plasticity: The Legacy of Graham Goddard*, ed. Morell, F., pp. 110–123. Massachusetts: Springer-Verlag.
14 McNaughton, B.L. and Morris, R.G.M. (1987). *Trends Neurosci.* **10**, 408–415.
15 Mizumori, S.J.Y., McNaughton, B.L., and Barnes, C.A. (1989). *J. Neurophysiol.* **61**, 15–31.
16 O'Keefe, J. and Dostrovsky, J. (1971). *Brain Res.* **31**, 171–175.
17 O'Keefe, J. and Nadel, L. (1978). *The Hippocampus as a Cognitive Map.* Oxford: Clarendon Press.
18 Poucet, B., Thinusblanc, C., and Muller, R.U. (1994). *Neuroreport* **5**, 2045–2048.
19 Quirk, G.J., Muller, R.U., Kubie, J.L., and Ranck, J.B., Jr. (1992). *J. Neurosci.* **12**, 1945–1963.
20 Ranck, J.B., Jr. (1973). *Exp. Neurol.* **41**, 461–531.
21 Rolls, E.T. (1989). In *Neural Models of Plasticity: Experimental and Theoretical Approaches*, ed. Byrne, J.H. and Berry, W.O., pp. 240–265. San Diego: Academic Press.
22 Rose, G., Diamond, D., and Lynch, G.S. (1983). *Brain Res.* **266**, 29–37.
23 Scoville, W.B. and Milner, B. (1957). *J. Neurol. Neurosurg. Psychiat.* **20**, 11–21.
24 Sharp, P.E. and Green, C. (1994). *J. Neurosci.* **14**, 2339–2356.
25 Sutherland, R.J., Wishaw, I.O., and Kolb, B. (1983). *Behav. Brain Res.* 7, 133–153.
26 Treves, A. and Rolls, E.T. (1991). *Network* **2**, 371–397.
27 Wilson, M.A. and McNaughton, B.L. (1993). *Science* **261**, 1055–1058.
28 Witter, M.P. (1986). In *Excitatory Amino Acids and Epilepsy (Adv. Exp. Med. Biol., vol. 203)*, ed. Schwarcz, R. and Ben-Ari, Y., pp. 67–82. New York: Plenum Press.

6

THE WARP AND WEFT IN THE INFERIOR TEMPORAL CORTEX

ICHIRO FUJITA

Department of Cognitive Neuroscience, Osaka University Medical School, and Precursory and Embryonic Science and Technology, Research Development Corporation of Japan, Suita, Osaka 565, Japan

The cerebral neocortex is not a randomly interconnected set of neurons, but a highly structured fabric woven from radial (vertical) and tangential (horizontal) connections. The radial arrangement of neurons and connections can be recognized throughout the entire neocortex. However, functionally defined columns (*i.e.*, clusters of neurons with similar response characteristics or functions across all or most cortical layers) have been demonstrated for only a few sensory cortical areas. In the primary visual cortex, neurons are organized into columnar organization according to their ocular dominance preference and orientation selectivity (*9*). The primary visual cortex is also the area in which intrinsic horizontal axons have been a major focus of current research (*7*). In association cortices, by contrast, evidence of functional columns is scarce, and the horizontal axon system has just begun to be demonstrated in the past few years (*1*, *20*).

The lack of evidence for functionally defined columns in association cortices may be partly because neurons upstream along the cortical hierarchy show highly integrated response properties, and their stimulus selectivity is difficult to characterize. The lack of evidence may also be due to the notion that association cortices are critically involved in dynamic cognitive functions, such as perception. Emphasis on the dynamic aspects of the neuronal network often discourages efforts to find organization such as columns which remain stable, at least for a certain period of time.

Cytoarchitectonic area TE of the inferior temporal cortex of the macaque monkey is a higher association cortex at or near the final stage of the visual pathway essential to object recognition. Lesions in this area severely impair the monkey's ability to discriminate shapes and recognize objects (*8*). Use of a computer graphics system in generation and presentation of visual stimuli has recently made it possible for visual stimulus to be generated and modified *on line* while activity of a single cell in the TE is monitored, allowing the minimal combination of stimulus features required to activate the cell to be identified. This method has led us to the discovery that the TE consists of columns, each containing cells responding to similar features of objects (*3–5*). In this article, I shall describe visual response characteristics of TE neurons and physiological aspects of the columnar organization of the TE. Local neuronal circuitry related to these functional columns, particularly the intrinsic horizontal axons, will then be described. Finally, I shall consider interactions among columns and how the TE operates in response to viewing objects.

I. RESPONSE PROPERTIES OF TE NEURONS

TE neurons have unique physiological characteristics compared to those in the preceding areas along the cortical pathway for object vision. When monkeys are passively shown various objects, TE cells respond to only a subset of them. The simplification procedure, in which stimulus features critical for eliciting responses are determined by step-by-step reduction of stimulus complexity using a computer graphics system (*4*), shows that TE neurons actually respond to visual features of objects such as a particular shape, color, texture or a combination of these, not to an object as a whole (*2–4*, *13*, *18*, but see Perrett *et al.* in this volume for face selective cells). Examples include a cross, a striped square, or a colored star. These stimuli are far more complex than those effectively evoking responses in the primary visual cortex such as oriented bars or colored spots, yet they are still simpler than objects in the real world. If two objects share a feature to which a TE cell is selective, then the cell responds to both of them (Fig. 1).

Receptive fields of TE cells are large: the square root of the receptive field area is 15–20 degrees on average (*2, 8, 13, 18*). Fovea is almost always included in the receptive field. Neurons in the TE have largely overlapping receptive fields, and the TE has no retinotopic map. TE neurons respond to a stimulus at any site within their receptive field, and stimulus selectivity is largely preserved across the receptive field; this property is referred to as translation invariance (*8, 11*). However, they are sensitive to changes in stimulus orientation (*18*). Changes of stimulus in other dimensions such as contrast polarity and size affect responses in half of the TE cells, but responses of the other half do not vary with these stimulus dimensions (*10, 11*).

Fig. 1. TE cells respond to visual features of objects, and do not signal abstract representation of objects *per se*. If a cell is selective for a feature shared by two objects, the cell responds to both of them. The cell shown here, for example, responded to a girl's face and a red pepper. The stimulus feature critical for activating this cell is a horizontally elongated shape with the upper half dark, lower half light, and the center portion darkest. This feature was demonstrated both by the girl's lips and a part of the pepper.

II. COLUMNAR ORGANIZATION IN TE

Stimulus features that elicit responses from TE neurons are quite diverse. The question is how diverse they are. Does each neuron have a preferred stimulus totally different from that preferred by other neurons? An earlier study (*18*) indicated that this is not the case; cells which are recorded at successive positions along an electrode penetration tend to respond to the same or similar stimuli, suggesting that TE cells with similar stimulus selectivity cluster together. Simultaneous recordings from adjacent TE neurons have confirmed this by showing that adjacent cells respond maximally or nearly maximally to the same stimuli (*3*). Systematic analysis of a series of neurons along vertical and tangential penetrations then have demonstrated that the cluster is columnar in shape (Fig. 2). In vertical penetrations, neurons distributed over nearly the entire thickness of the gray matter shared the optimal or near-optimal stimuli. Neurons with similar selectivities are thus arrayed vertically across most layers of the cortex. In tangential penetrations, cells responding to similar stimuli are localized in patches 0.4 ± 0.2 mm wide. Cells immediately outside the cluster did not respond to stimuli which activated cells in the cluster. When we succeeded in determining effective stimuli for these unresponsive cells outside the first cluster, the stimulus features

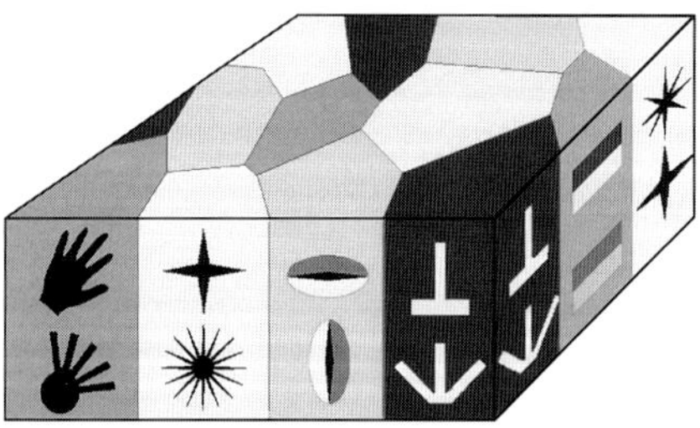

Fig. 2. Functional columnar organization in area TE. Neurons with similar stimulus preferences group together to form columns 0.4 ± 0.2 mm wide. Neurons within a column responded to the same stimuli, but the exact optimal stimulus or tuning properties differed from each other, as exemplified by similar but different icons within a column. Multiple columns were selective for similar features, and are shown in the same shading patterns. See text for further details.

were distinctly different from those of the first cluster. The border between two adjacent columnar clusters is rather discrete. When a recording electrode was moved across the border by 50 μm we noticed a drastic change of effective stimulus features (*5*).

There appear to be multiple columns which respond to similar stimulus features, because we found two or three clusters of neurons responding to the same set of stimuli along a tangential penetration 3–4 mm long. Exactly how these multiple columns are similar to or different from each other, however, awaits further studies. Another important characteristic of columns in the TE is that neurons in a single column show some variances in their stimulus selectivity. For example, in a column where neurons respond to star shapes, some prefer stars with 4 protrusions, while other neurons prefer stars with 16 protrusions. This kind of difference in stimulus selectivity is found even between 2 adjacent neurons within a column.

Preliminary experiments suggest that the stimulus preference of columns remains stable at least over 30 days, because repeated recordings from the same site over this period allowed us to find neurons with similar stimulus preferences (*5*). This, however, does not exclude the possibility that neurons in the TE can change their stimulus selectivities. Indeed, it was recently shown that TE neurons change their stimulus selectivity as a result of learning (*12*). Monkeys were trained to perform a delayed matching-to-sample task in which they had to retain the memory of an artificially synthesized visual stimulus and, after a delay, discriminate it from other stimuli. After monkeys had established a success rate of more than 85% with 28 visual stimuli, the TE of the trained monkeys was surveyed under anesthetized conditions. A substantial proportion (40%) of neurons in the trained monkeys responded to the trained stimuli, whereas only 9% of TE neurons responded in naive monkeys. The results, taken together, suggest that the columnar organization in the TE is dynamically regulated and maintained, and can undergo changes when necessary.

The discreteness and temporal stability of columns provided us with the rationale to estimate the number of columns in the TE. If we take a square of the columnar width (0.4 mm) measured in physiological experiments as the area of an individual column (0.16 mm^2)

and divide the surface area (330 mm^2) of the dorsal part of the TE where we made recordings, the number of columns is estimated to be 2,000 (*3*). If we calculate the number on the basis of width of intrinsic and afferent fiber terminal patches (0.5 mm, see below), the number would be 1,300.

III. LOCAL NEURONAL CIRCUIT IN THE TE

Some features of local circuits in the TE may be related to the functional columnar organization of this area. The TE contains extensive vertical connections between the cortical layers as well as intrinsic horizontal axons running parallel to the pial surface (Fig. 3) (*6*). Horizontal axons run for a length of more than 4 mm, sometimes up to 8 mm, in the TE, and produce plexuses of terminal arborization at certain intervals. Terminal plexuses form ovoid shapes across the cortical surface and are columnar along the plane perpendicular to the surface. Axons and terminal boutons in the plexuses are mainly distributed from layer 1 to 3, and are sometimes found from layer 1 to 5. Each plexus measures 0.5 mm wide on average which coincides with the size of functional columns. The distance between columns with similar selectivity (0.4–1.0 mm) (*3*) is comparable to the range of distances between plexuses of horizontal axons. Terminal arborization of afferent fibers from the TEO, which is one step earlier in the object vision pathway, also takes a columnar form of similar size in

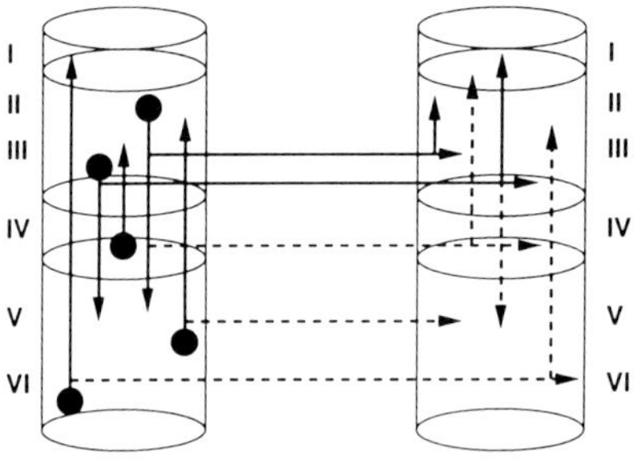

Fig. 3. Schematic diagram of local neuronal connections within and between columns. Black circles denote cell bodies. Arrows with solid lines indicate strong projections and arrows with dashed lines weaker projections. The connections so far confirmed are shown (*6*).

the TE (*16*). All these anatomical features, vertical interlaminar connections and columnar arborization of intrinsic horizontal axons and afferent fibers, make extensive interactions across layers possible, and can contribute to the shared stimulus selectivity among neurons within a column.

IV. INTERACTIONS MEDIATED BY HORIZONTAL AXONS

Intrinsic horizontal axons enable distant columns to interact with each other. Horizontal axons in the TE, however, are not for interactions between neurons responding to different parts of the visual field as proposed for the primary visual cortex, because TE neurons have overlapping receptive fields. It is likely that horizontal connections are related to the stimulus selectivity of columns. It is tempting to assume a relation between patches of horizontal axon terminals and multiple columns with similar selectivity, but no attempt has been made to correlate the anatomy and physiology. Do horizontal axons link columns with similar stimulus preferences or do they link columns with different stimulus selectivities? The answer to this question will be a key step in understanding what information horizontal axons convey and whether horizontal axons contribute to the generation of complex stimulus selectivity or formation of a large receptive field or serve totally different functions.

Experiments in which responses to two simultaneously presented stimuli are examined may be relevant to the question of interactions between columns. When two stimuli, either identical or different, are presented in the receptive field of neurons in the TE and the lower bank of the superior temporal cortex, response never becomes a summation of responses obtained with each stimulus presented separately (*15, 17*). Responses to two stimuli shown together often become weaker than those to single stimuli. Because this suppression occurs even when two stimuli are separated by more than 10°(*17*) and is independent of whether two stimuli are identical or different (*15*), it has been suggested that this suppressive interaction occurs in the TE and the lower bank of the superior temporal sulcus, not in the earlier stages such as V4. If one determines the critical stimulus features for a cell under study and carefully chooses stimuli for 2 stimuli experiments, interactions between columns can be probed further.

Identification of the cells of origin of horizontal axons and their postsynaptic targets is a necessary step toward understanding interactions among columns. A single injection of a fluorescent retrograde tracer, diamidino yellow, into the TE, which covers all layers across an area 0.2–0.4 mm in tangential width, will label more than 20 columnar clusters of cells as well as diffusely distributed cells within the TE. In the clusters, labeled cells are found from layer 2 to 6 with no or significantly fewer labeled cells in layer 4 than in the other layers. These cells are the cells of origin of horizontal axons which take up the diamidino yellow at the injection site. Intracellular injections of another fluorescent dye, Lucifer Yellow, into these cells have revealed that the vast majority of cells of origin of horizontal axons in the TE are pyramidal cells both in supragranular and infragranular layers (*19*). Even the rare cells labeled in layer 4 show pyramidal cell morphology. Thus, most horizontal axons are excitatory in their synaptic transmission. Cells postsynaptic to horizontal axons have not been identified in the TE. In the cat primary visual cortex, 20% of the cells which receive synaptic contacts from horizontal axons are smooth stellate cells, *i.e.*, inhibitory interneurons (*14*).

V. WHEN A MONKEY SEES AN OBJECT

The findings described above allow some speculation on how the TE responds to an object. When an object is presented to central visual field within 10–20°, a particular subset of columns in the TE will be activated. A single feature in the visual image of the object activates multiple columns which are selective for that feature. Ordinary objects have rich visual features in their images and each feature activates a different set of columns. In theory, the combination of these active columns can represent the composition of the entire image of the object. Visually different objects activate different sets of columns, and similar objects activate partially overlapping sets of columns. The principle of combination equips the TE with a capacity to represent a huge number of objects, much like the way in which 26 letters in the alphabet produce hundreds of thousands of words and an infinite number of sentences ("*Visual alphabet hypothesis*") (*4*).

Although the analogy to the alphabet explains the representation capacity of columns in the TE, some functional aspects of columns in the TE are not consistent with the analogy. First, activity in each column can signal more than "yes" or "no" regarding the presence of a particular feature in an object. As noted earlier, neurons in a column have similar, but not identical, stimulus selectivities. A slight change in a visual feature due to change in the viewing angle or illumination activates overlapping but different populations of neurons in a column that is selective for that feature. The activity pattern across a population of neurons within a column can thus signal slight changes in visual images. The second deviation from the alphabet analogy is that the combined representation in the TE may not be a simple addition of active columns. Let us suppose that feature A activates a group of columns A′ and feature B activates a group of columns B′. The visual alphabet hypothesis predicts that when features A and B are combined and presented, both groups of columns, A′ and B′, will be activated. However, the activity level in columns A′ and B′ caused by a combination of features A and B may not be the same as that when each feature is presented separately. Horizontal interaction between columns exerts a modulation of the activity of both columns when two features are shown in combination, as suggested by experiments involving simultaneous presentation of two stimuli. Finally, the alphabet should be common to all members in the same language group, because it is used for communication among group members. However, stimulus features represented in columns can differ from individual to individual, since they are used only within the individual brain. When we consider the plasticity of TE neurons (*12*), individuals with different life experiences may well have a different set of stimulus features represented in columns within the TE. However, I should add that some stimuli such as star shapes, T shapes, and concentric rings were effective stimuli for certain TE neurons across individual animals, suggesting that some stimulus features represented in the TE are common to all monkeys.

SUMMARY

Area TE of the inferior temporal cortex, the final or near-final stage of the monkey visual cortical pathway crucial for object recognition, consists of columns. TE cells with similar selectivity cluster in multiple columns separate from those with different stimulus preferences. The width of each column is 0.4–0.5 mm on average. Repeated recordings from the same site suggest that stimulus selectivity in the columns is maintained for at least 30 days. The discreteness and temporal stability of TE columns provided us with the rationale for estimating the number of columns in area TE. The estimated number was 1,300–2,000. Ordinary objects contain rich visual features in their images and each feature activates a different set of columns. In theory, the combination of these active columns represents the composition of the entire image of the object. However, the TE contains an extensive network of intrinsic horizontal axons which mediate complex interactions among columns. The combined coding of object images, therefore, may not be a simple linear addition of active columns.

Acknowledgment

Part of the studies reviewed here was supported by grants from the Uehara Memorial Foundation, the Casio Science Foundation, and the Ichiro Kanehara Foundation.

REFERENCES

1 Amir, Y., Harel, M., and Malach, R. (1993). *J. Comp. Neurol.* **334**, 19–46.
2 Desimone, R., Albright, T.D., Gross, C.G., and Bruce, C. (1984). *J. Neurosci.* **4**, 2051–2062.
3 Fujita, I., Tanaka, K., Ito, M., and Cheng, K. (1992). *Nature* **360**, 343–346.
4 Fujita, I. (1993a). *Biomed. Res.* **14**, Suppl. 4, 21–27.
5 Fujita, I. (1993b). *Neurosci. Res. Suppl.* **18**, S13.
6 Fujita, I. and Fujita, T. (1993). *Abstr. Soc. Neurosci.* **19**, 971.
7 Gilbert, C.D. (1993). *Cerebral Cortex* **3**, 373–386.
8 Gross, C.G. (1973). In *Handbook of Sensory Physiology*, ed. Jung, R., pp. 451–482. Berlin: Springer.

9 Hubel, D.H. and Wiesel, T.N. (1962). *J. Physiol.* **160**, 106–154.
10 Ito, M., Fujita, I., Tamura, H., and Tanaka, K. (1994). *Cerebral Cortex* **5**, 499–508.
11 Ito, M., Tamura, H., Fujita, I., and Tanaka, K. (1995). *J. Neurophysiol.* **73**, 218–226.
12 Kobatake, E., Tanaka, K., Wang, G., and Tamori, Y. (1993). *Soc. Neurosci. Abstr.* **19**, 975.
13 Kobatake, E. and Tanaka, K. (1994). *J. Neurophysiol.* **71**, 856–867.
14 McGuire, B.A., Gilbert, C.D., Rivlin, P.K., and Wiesel, T.N. (1991). *J. Comp. Neurol.* **305**, 370–392.
15 Miller, E.K., Gochin, P.M., and Gross, C.G. (1993). *Brain Res.* **616**, 25–29.
16 Saleem, K.S., Tanaka, K., and Rockland, K.S. (1993). *Cerebral Cortex* **3**, 454–464.
17 Sato, T. (1989). *Exp. Brain Res.* **77**, 23–30.
18 Tanaka, K., Saito, H-A., Fukada, Y., and Moriya, M. (1991). *J. Neurophysiol.* **66**, 170–189.
19 Tanigawa, H., Fujita, I., Ojima, H., and Kato, M. (1995). *Proc. 4th IBRO Congress*, in press.
20 Yoshioka, T., Levitt, J.B., and Lund, J.S. (1992). *J. Neurosci.* **12**, 2785–2802.

7
ACQUIRED FORM REPRESENTATION IN VISION

KUNIYOSHI SAKAI*

Department of Physiology, School of Medicine, University of Tokyo, Tokyo 113, Japan

Primate vision has a remarkable ability to recognize a variety of similar faces or objects. This ability suggests that form information is perceptually organized so that it enables fine discrimination of faces or objects. In this review I use the term *form* for the geometry of an object's overall structure. I believe that form is an indispensable concept in understanding mechanisms of object recognition, because form directly represents an object's entity and enables its recognition.

The inferotemporal (IT) cortex has been proposed to be the memory storehouse in object vision (*9, 15, 24, 26–28*). Along the visual pathway from the primary visual cortex (V1) to the anterior inferotemporal (AIT) cortex, both the receptive field size and the complexity of neuronal processing increase (*10, 33*). Consequently, IT neurons respond selectively to complex forms such as hands, faces, and computer-generated forms (Fig. 1). Whereas the orientation selectivity of V1 and V2 neurons has been well characterized, the form selectivity of IT cells has not been thoroughly studied, owing in

*Present address: MGH-NHR Center, Massachusetts General Hospital & Harvard Medical School, Charlestown, MA02129, U.S.A.

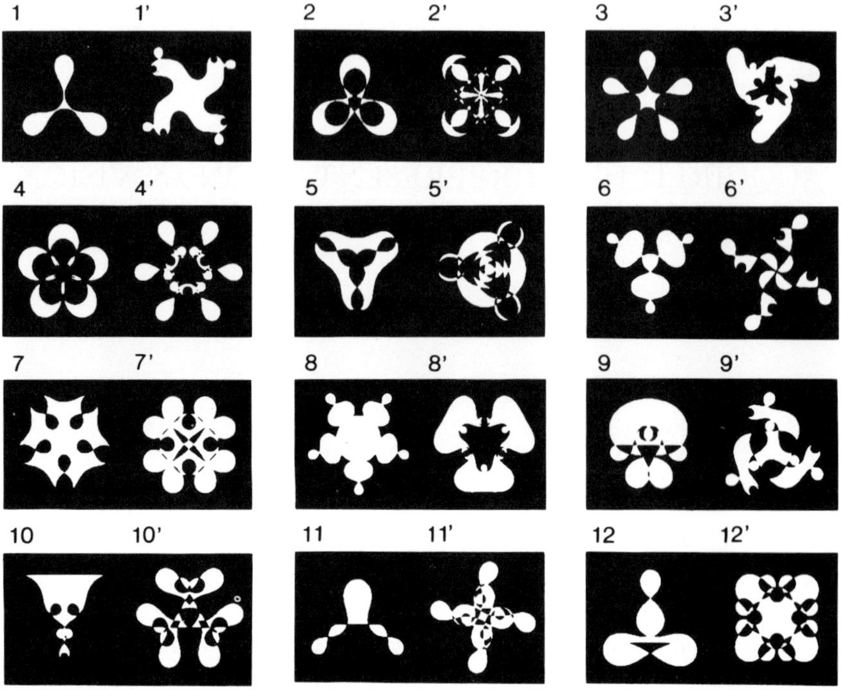

Fig. 1. Twelve pairs of Fourier descriptors (1 & 1′ to 12 & 12′) for stimuli in the pair-association task. When one member of each pair is shown, trained monkeys can retrieve and select the other member of the paired associates.

part to the complexity of form features. When recording a V1 cell, its orientation preference and selectivity can be systematically determined. Orientation is of one-dimensional parameter ranging from 0° to 180°. In contrast, the range of form variety is almost infinite. Moreover, no critical parameters are known for the specification of general forms.

There still remains a controversial question whether form perception occurs as overall identification or as a synthesis of structural components. Some theories of object recognition have proposed bottom-up determination of an object's components and subsequent matching of the arrangement of components with a memory representation (*2*, *13*, *20*). The properties of face-selective cells may be in agreement with these theories (*21*). However, there is psychological evidence that object features (not specifically related to parts) are matched directly with such overall features stored in long-term memory (*5*).

If the form analysis in vision is hierarchical and bottom-up, then some aspects of the form selectivity of IT cells reflect the constraint in the early visual processing. Besides, if this visual processing is based on computational logic, then there must be some principles that guide the generation of the form selectivity. In this review I propose the guiding principles of association and neuronal tuning in the AIT cortex. My working hypothesis is that learning of a form or repeated exposures to a form produces neurons that selectively respond to that particular form. This idea bears some relation to prior conjectures for the picture-selective responses that depend on temporal contiguity (*16*).

I. PAIR-ASSOCIATION TASK

Most of our long-term memories of episodes or objects are organized so that we can retrieve them by association. Anderson and Bower (*1*) proposed that human memory only stores "propositions", which are conceived as "structured bundles of associations between elementary ideas or concepts". This notion may be applied to object recognition: visual memory stores forms as structured bundles of associations between elementary views of objects. The neurophysiological evi-

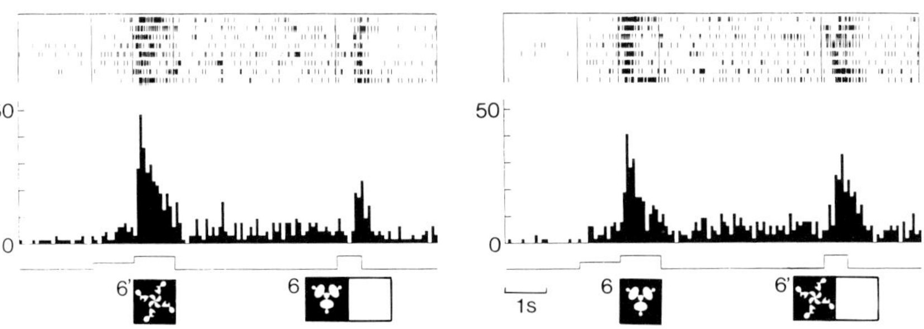

Fig. 2. Responses of a pair-coding neuron, which exhibited form-selective activity during the cue period. Left: trials for cue 6′ that elicited the strongest cue response. Right: trials for cue 6 that elicited the second-strongest cue response.

dence of associative mechanisms in visual memory has been found only recently (*23, 24*).

Figure 1 shows a set of visual stimuli that I have used. These pictures were sorted randomly into pairs, numbered from 1 & 1′ to 12 & 12′. The pair combinations were fixed throughout the experiments. To impose the acquisition of this screening set (see the section IV) as long-term memory, I trained monkeys (*Macaca fuscata*) in the pair-association (PA) task (*18, 23*). In each trial of the PA task, a cue stimulus (one of 24 pictures shown in Fig. 1) is presented at the center of the video monitor for 0.5 sec. After a delay period of 5 sec, two stimuli, the paired associate of the cue (correct choice) and a distractor (incorrect choice) are shown. The monkey obtains a fruit juice reward for touching the correct one within 1.2 sec. This task paradigm can reliably demand the learning of visual stimuli, because monkeys cannot select a paired associate correctly without memorizing and recalling pair combinations.

It should be noted that the PA task cannot be solved by employing short-term or working memory within a single trial. Instead, the PA task is essentially the memory *recall* task, which explicitly demands the memory retrieval and thus generation of images from the long-term memory. Therefore, memory components of the PA task present a sharp contrast to those of the delayed matching-to-sample (DMS) task. In the DMS task, the subject indicates whether a test stimulus matches a previously shown sample stimulus. When the stimulus set size is small and thus the same trial is repeated, working memory is mainly involved in the DMS task. When the stimulus set size is large and thus each trial is unique, recognition memory is also involved, because it is possible to indicate whether a test stimulus has appeared or not without employing working memory. We have to take these points into consideration in interpreting the results of behavioral and physiological studies.

II. PAIR-CODING NEURON

In the AIT cortex, I found one type of neuron (*pair-coding neuron*), which manifested selective responses to both paired associates (*23*). The properties of pair-coding neurons indicate that memory storage is organized such that single neurons can code both paired associates in the PA task (Fig. 2). This result provides new evidence that single neurons acquire form selectivity through associative learning. Here, this type of coding is termed *associative coding*, in which the involvement of associative learning is essential for memory storage.

The associative coding proposed here provides one organizing principle by which the special selectivity of neuronal responses is produced. The spatiotemporal patterns of neuronal discharges selective to object forms thus constitute the basis of ensemble coding. A possible molecular mechanism of the associative coding lies in the change of synaptic connections through repetitive learning, whereby two inputs are always paired with each other. The associative mechanism based on temporal contiguity is further discussed in section VIII.

III. PAIR-RECALL NEURON

I found another type of neuron (*pair-recall neuron*) that is presumably involved in the process of memory retrieval (*23*). In the PA task, the monkey is required to recognize the paired associate of a cue stimulus after a delay period. Pair-recall neurons exhibited form-selective delay activity (Fig. 3). This response is closely coupled with the paired

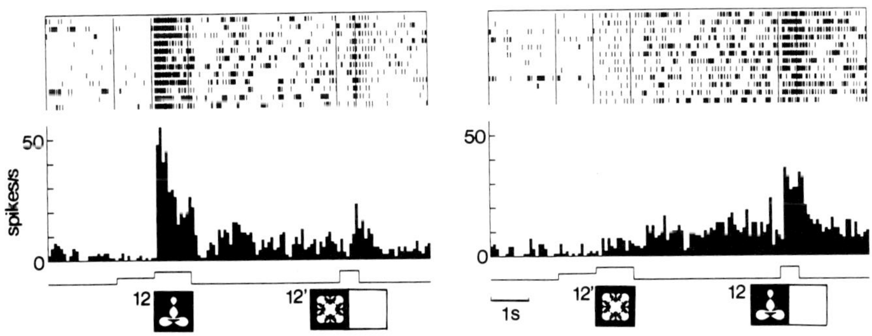

Fig. 3. Responses of a pair-recall neuron, which exhibited form-selective activity during the delay period. Left: trials for cue 12 that elicited the strongest cue response. Right: trials for cue 12′ that elicited the delay response, presumably reflecting retrieval of the paired associate 12. Note the tonic increasing activity during the delay period, which is much stronger than the cue response.

associate that is not actually seen, but retrieved by the cue stimulus. There are two possibilities for the critical process during the delay period. One is to hold a *retrospective code*, which is a cue stimulus, in working memory. The other is to generate a *prospective code* by converting a cue into its paired associate. The increasing delay activity of the pair-recall neurons is consistent with the claim that subjects can employ a prospective code. On the grounds that the AIT cortex serves as the memory storehouse, these neurons could serve as memory storage elements, also activated in the retrieval process. The finding of pair-recall neurons is the first neurophysiological demonstration that visual imagery is also implemented by the same neuronal mechanism that subserves memory retrieval.

IV. FINE-FORM SELECTIVITY

The conventional method for determining the response selectivity of a single IT neuron utilizes a *screening set* of various object forms. Because experimenters are not able to test every possible form, they must prepare a convenient screening set with many visual stimuli. Figure 1 shows an example of a screening set that can specify global form selectivity. If a recorded neuron responds to at least one of the forms in a screening set (responsive), but not equally to all of them (selective), the global form selectivity of this cell can be further characterized.

One may mistakenly conclude that the most effective form in a screening set is the optimum stimulus for a recorded cell. One way to overcome the limitation of using a screening set is the analysis of *fine-form selectivity*, which provides better resolution to discriminate among effective stimuli with similar features. In other words, the test of a cell's form selectivity should be performed in two steps. First, test the cell's responses with a broad screening set. Second, test them more finely by preparing forms similar to the effective stimuli selected in the first step. The analysis of fine-form selectivity becomes particularly important in searching for memory traces of specific forms, which could be acquired through learning experience.

V. EVIDENCE OF NEURONAL TUNING

The patterns shown in Fig. 1 are Fourier descriptors (FDs). FDs are appropriate in testing fine-form selectivity because they are specified by a set of parameters: harmonic amplitude (A_k) and phase angle (α_k), where $k = 1,2,...$ is each term's ordinal number in a Fourier series (*34*). A slight alteration of one FD parameter from its original value produces a very similar form. This manipulation of an original pattern is called *parametric transformation.* The similarity of forms with slightly altered FD parameters is validated by the fact that the position of each point in the drawing plane is a continuous function of FD parameters (some examples are shown in Fig. 4).

Figure 4 shows data from one AIT neuron with form-selective

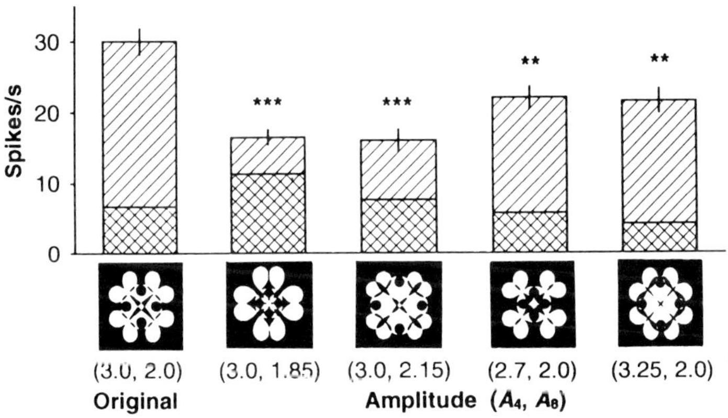

Fig. 4. Responses of an exemplary neuron for the parametric transformation. The original pattern shown is specified by two harmonic amplitudes and two phase angles ($A_4 = 3.0$, $\alpha_4 = \pi/2$, $A_8 = 2.0$, $\alpha_8 = \pi/2$). The set of two numbers attached to each pattern denote its amplitude parameters (A_4, A_8). In each transformed pattern, one of these parameters is slightly altered from the original value. Note the close similarity between these four transformed patterns and the original. Histograms show mean discharge rates (mean $\pm$ S.E.M.) for each cue presentation of the original or transformed patterns. The cross-hatched area in each histogram bar corresponds to the spontaneous discharge rate. This neuron exhibited optimal responses to the original pattern rather than to the transformed patterns. These data were taken from the first five trials in which the original or transformed pattern was used as a cue stimulus. Two asterisks, $p < 0.01$; three asterisks, $p < 0.001$: according to t test with unequal variances (*29*).

responses. The critical comparison is the neuronal response to the learned form *versus* unlearned transforms during the cue period. The original FD pattern shown in Fig. 4 has two amplitude parameters. A_4 and A_8 that can take continuous values. It can be readily noted that the global features of patterns with a slightly altered value of A_4 or A_8 are strikingly similar. All of the transformed patterns elicited significantly weaker responses than the original pattern. This observation was further confirmed by responses of other neurons. The parametric transformation resulted in a significantly weaker neuronal response for most cases. Moreover, I did not observe any case in which responses to transformed patterns were stronger than those to the original patterns (*25*, *28*). This result suggests that AIT neurons are subject to *tuning* mechanisms for particular forms in a learning process. This neuronal tuning cannot be explained by learning-independent innate selectivity, since the original patterns in the screening set were selected randomly.

For the following reasons, the possibility that the weaker response to unlearned transforms is attributable to purely sensory reasons can be excluded. First, every tested transform was derived from one of the original patterns, which elicited the strongest or the second-strongest response. Therefore, these transforms are likely to produce equally strong responses as the original pattern, obviously stronger than other ineffective original patterns. Furthermore, it is impossible to predict beforehand the relative effectiveness between the original pattern and transforms that derived from it, because original patterns have no special form attributes and their FD parameters were selected randomly. Second, a cell's preference within the screening set should not be confused with fine-form selectivity among transforms, because the range of form variety in the screening set is incomparably wider than that in transforms. Therefore, the learning-dependent *fine*-form selectivity cannot be explained by the *global* form selectivity for the screening set.

VI. INTERACTION BETWEEN MEMORY MECHANISMS

Pair-coding neurons and pair-recall neurons can participate, respectively, in the coding and recall processes of visual long-term memory.

The pair-recall neuron represents a typical example of interaction between sustained activation and association among the neuronal mechanisms that subserve the formation or expression of memory traces (*6*). Evidence for form-selective or color-selective sustained activation has been documented in the IT cortex (*7, 8, 17*). Furthermore, there is another type of interaction among neuronal mechanisms. The fact that each pair-coding neuron responds to both paired associates suggests an important role of interaction between neuronal tuning and associative mechanisms.

VII. IS FORM REPRESENTATION 2D VIEWER-CENTERED OR 3D OBJECT-CENTERED?

One of the most fundamental problems in object recognition is how a 3D object form in the real world is represented in long-term memory. Marr and Nishihara (*14*) proposed that a form representation for recognition should use a 3D *object-centered* coordinate system, which is transformed from a 2D *viewer-centered* coordinate system. In a viewer-centered coordinate system an object's geometry is specified relative to the viewer. In an object-centered coordinate system an object's geometry is specified relative to the object itself. The latter emphasizes the computation of a form-specific description that is independent of the vantage point (*2*). In contrast to this 3D object-centered model, recent studies propose that multiple 2D views are directly stored for each object representation (*3, 4, 12, 22, 31*). In this scheme, a viewer-centered recognition can be achieved by interpolating between a small number of stored views. To establish that memory representations of object forms are multiple 2D views rather than internal 3D models, we need more supporting data that can elucidate the underlying processes for object recognition.

VIII. THE IMPLICIT ASSOCIATIVE PROCESS AND OBJECT RECOGNITION

I further propose a possibility that associative mechanisms are shared by two neuronal processes: one is the explicit process that is critical in the PA task, whereas the other is the implicit or automatic process. The latter implicit associative process may subserve object recognition, in which distinct views of an object can be treated as common views of the same object because these views are overwritten in the common neuronal network that consists of many pair-coding neurons. Using this scheme, one standard view can be naturally associated with any observed views of the same object, thus enabling correct recognition.

In our visual world, multiple views of an object are nearly always presented in succession, resulting from relative movement between the observer's eyes and the object. This situation is an example of the implicit associative process in that two different views of an object are associated and memorized by AIT neurons. Such an associative mechanism based on temporal contiguity agrees well with the principle of generic image of sampling proposed by Nakayama and Shimojo (*19*). Incorporating this significant conceptual framework, the implicit associative process can be elucidated further: the relationship of generic 2D views is automatically acquired by the neuronal mechanism of association.

IX. A MODEL OF THE COGNITIVE MEMORY SYSTEM

A model of the cognitive memory system that I have proposed (*24, 26, 28*) is shown in Fig. 5A. This scheme is based on structures and functions of the visual memory system for object recognition (Fig. 5B), and it contains some ideas that have been put forward by several researchers (*9, 15, 30, 32, 33*).

In the memory acquisition process, sensory information is transformed into a memory code of neuronal responses (encoding) with the bottom-up information flow from feature analyzers to a memory storehouse. The primary visual area and the prestriate area serve as feature analyzers in vision. A possible candidate for the memory storehouse is the temporal association area. In visual perception, prominent features in a visual field are selected and located by a focal-attention controller. One candidate for the focal-attention controller is the pulvinar nucleus. I hypothesize that perception is implemented by the interaction between memory acquisition (encoding) and focal-attention mechanisms (Fig. 5A).

To establish a long-lasting representation in the memory store-

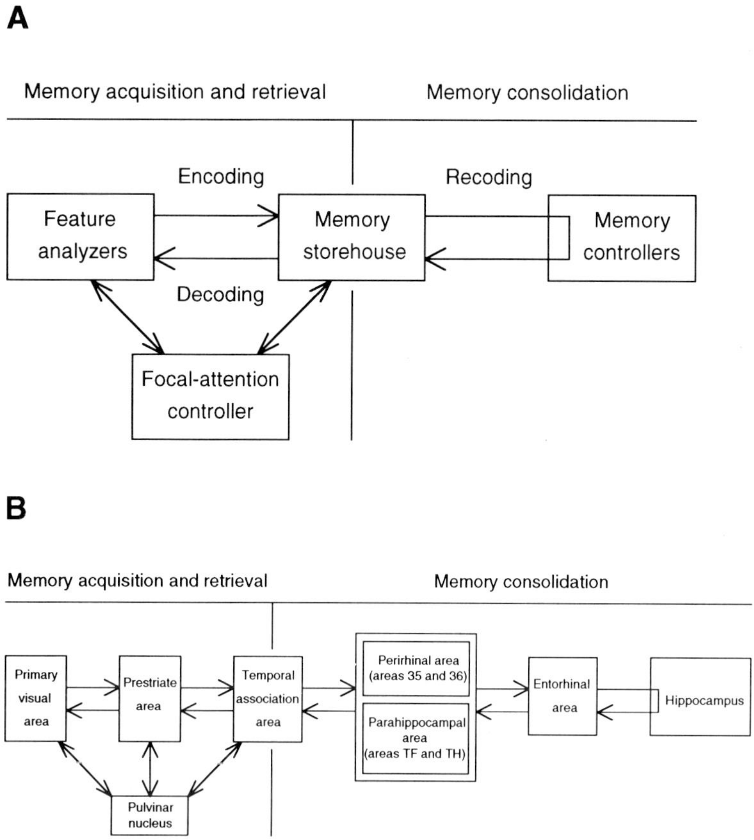

Fig. 5. A: a model of the cognitive memory system that unifies perception and imagery. Perception is implemented by the interaction between memory acquisition (encoding) and focal attention mechanisms. Imagery is implemented by the interaction between memory retrieval (decoding) and focal attention mechanisms. B: structures of the visual memory system for object recognition. Information flows between these structures correspond to the information flows indicated in A. Brodmann's area numbers 35 and 36, and von Economo's area symbols 'TF' and 'TH' are indicated. Note that most of the pathways between cortical visual areas are anatomically reciprocal, supporting their functional roles in encoding and decoding processes.

house, memory controllers are responsible for consolidation. During this memory consolidation process, the memory code is reorganized dynamically (recoding) by the interaction between the memory storehouse and memory controllers. Medial temporal structures, including the hippocampus, are regarded as memory controllers, based on their plasticity (*e.g.*, long-term potentiation and depression), neural networks (*e.g.*, autoassociation), clinical observations (*e.g.*, medial temporal lobe amnesia), and lesion studies.

In the memory retrieval process, memory codes stored in the memory storehouse are decomposed into more elementary attributes (decoding) by feature analyzers through a back-projection pathway. I hypothesize that imagery is implemented by the interaction between memory retrieval (decoding) and focal-attention mechanisms. As Kosslyn (*11*) pointed out, "imagery consists of brain states like those that arise during perception but occurs in the absence of the appropriate immediate sensory input". According to my scheme, visual imagery is generated by top-down activation of perceptual representations that are selected and located by the focal-attention controller. It is also possible that a signal from the focal-attention controller gates the top-down information flow toward early visual areas that function in the decoding process. This model predicts that the extent of visual areas devoted to decoding mental images is controlled dynamically by focal attention. The supporting evidence for this proposal on visual imagery has been discussed elsewhere (*26*).

SUMMARY

I examine the hypothesis that the form representation in the AIT cortex is acquired through learning. According to this hypothesis, perceptual aspects of the temporal association area are closely related to its visual representation, in that the response selectivity of AIT neurons can be influenced by visual experience. On the basis of neurophysiological evidence, I summarize two neuronal mechanisms that subserve the acquisition of form selectivity in AIT neurons. The first mechanism is association, with which relevant pieces of visual information are stored together and retrieved from each other. The second mechanism is neuronal tuning to particular stimuli that were learned in a cognitive task. The acquired form selectivity is a key feature in the capacity of temporal cortical neurons to establish the form representation with multiple 2D views. On the grounds that long-term memory of objects is acquired and organized by at least these two neuronal mechanisms in the temporal association area, I further present a model of the cognitive memory system that unifies perception and imagery.

REFERENCES

1 Anderson, J.A. and Bower, G.H. (1980). *Human Associative Memory: A Brief Edition*, Hillsdale: Lawrence Erlbaum.
2 Biederman, I. (1987). *Psychol. Rev.* **94**, 115–147.
3 Bülthoff, H.H. and Edelman, S. (1992). *Proc. Natl. Acad. Sci. U.S.A.* **89**, 60–64.
4 Cavanagh, P. (1991). In *Representations of Vision: Trends and Tacit Assumptions in Vision Research*, ed. Gorea, A., pp. 295–304. Cambridge, U.K.: Cambridge University Press.
5 Cave, C.B. and Kosslyn, S.M. (1993). *Perception* **22**, 229–248.
6 Desimone, R. (1992). *Science* **258**, 245–246.
7 Fuster, J.M. (1990). *J. Neurophysiol.* **64**, 681–697.
8 Fuster, J.M. and Jervey, J.P. (1982). *J. Neurosci.* **2**, 361–375.
9 Gross, C.G. (1973). In *Handbook of Sensory Physiology*, ed. Jung, R., vol. VII/3, pp. 451–482. Berlin: Springer-Verlag.
10 Gross, C.G. (1992). *Philos. Trans. R. Soc. Lond. B* **335**, 3–10.
11 Kosslyn, S.M. (1988). *Science* **240**, 1621–1626.
12 Logothetis, N.K., Pauls, J., Bülthoff, H.H., and Poggio, T. (1994). *Curr. Biol.* **4**, 401–414.
13 Marr, D. (1982). *Vision*, New York. Freeman.
14 Marr, D. and Nishihara, H.K. (1978). *Proc. R. Soc. Lond. B* **200**, 269–294.
15 Mishkin, M. (1982). *Philos. Trans. R. Soc. Lond. B* **298**, 85–95.
16 Miyashita, Y. (1988). *Nature* **335**, 817–820.
17 Miyashita, Y. and Chang, H.S. (1988). *Nature* **331**, 68–70.
18 Murray, E.A., Gaffan, D., and Mishkin, M. (1993). *J. Neurosci.* **13**, 4549–4561.
19 Nakayama, K. and Shimojo, S. (1992). *Science* **257**, 1357–1363.
20 Pentland, A.P. (1986). *Artif. Intell.* **28**, 293–331.
21 Perrett, D.I., Mistlin, A.J., and Chitty, A.J. (1987). *Trends Neurosci.* **10**, 358–364.
22 Poggio, T. and Edelman, S. (1990). *Nature* **343**, 263–266.
23 Sakai, K. and Miyashita, Y. (1991). *Nature* **354**, 152–155.
24 Sakai, K. and Miyashita, Y. (1993). *Curr. Opin. Neurobiol.* **3**, 166–170.
25 Sakai, K. and Miyashita, Y. (1994a). *NeuroReport* **5**, 829–832.
26 Sakai, K. and Miyashita, Y. (1994b). *Trends Neurosci.* **17**, 287–289.
27 Sakai, K. and Miyashita, Y. (1994c). *Trends Neurosci.* **17**, 513–514.
28 Sakai, K., Naya, Y., and Miyashita, Y. (1994). *Learning and Memory* **1**, 83–105.
29 Snedecor, G.W. and Cochran, W.G. (1989). *Statistical Methods*, Ames: Iowa State University Press.
30 Squire, L.R. and Zola-Morgan, S. (1991). *Science* **253**, 1380–1386.

31 Ullman, S. and Basri, R. (1989). *Artif. Intell. Lab. Memo No. 1152*, Cambridge, MA.: MIT Artificial Intelligence Laboratory.
32 Ungerleider, L.G. and Haxby, J.V. (1994). *Curr. Opin. Neurobiol*. **4**, 157–165.
33 Van Essen, D.C., Anderson, C.H., and Felleman, D.J. (1992). *Science* **255**, 419–423.
34 Zahn, C.T. and Roskies, R.Z. (1972). *IEEE Trans. Comput.* **c-21**, 269–281.

8
CATEGORY LEARNING IN AMNESIA

BARBARA KNOWLTON

Department of Psychiatry, University of California, San Diego, CA92093, U.S.A.

The question of how people learn categories has been a popular topic for psychologists. In order to learn to categorize stimuli, a subject must abstract what is common about the stimuli as a group. This common information could take the form of a set of rules that define category membership, or knowledge of what a particularly good example of the category might look like. For example, when one listens to several pieces of music by the same composer, one might learn about each individual piece, but in addition, one might also learn something general about the composer's style, and this information could be used to make a decision about whether a new piece of music belongs to the same composer. By this view, two kinds of information are being learned in parallel. Information about individual examples of a category, and general information about the category would be learned by different memory systems, and hence, by different brain systems (*2*, *16*). However, there is an alternative view of category learning. Perhaps people are not learning two kinds of information in parallel, but rather they are basing category level judgments on knowledge of individual examples (*3*, *11*). One might decide that an item belongs in a particular category because it resembles a number of other items stored in memory that belong in the same category. For example, one might decide that a particular piece of music was the work of a certain composer because it is similar to a number of other pieces by the same composer stored in memory. By this second view, category knowledge has no special status because memory for items is adequate to account for categorization, and as such, performance may depend on only a single brain system.

In general, it has been difficult to distinguish between these two views because they are both plausible. However, the study of neuropsychological patients has begun to reveal clues as to how the brain accomplishes category learning. Amnesic patients, who have sustained damage to structures in the medial temporal lobe, including the hippocampus, or to midline diencephalic nuclei, exhibit a severe deficit in recall and recognition about facts and events (*10, 20*). This type of memory has been termed "explicit" or "declarative" (*1*). The memory deficit in amnesic patients extends to all modalities, and is enduring. Although amnesia can result from damage to either the medial temporal lobe or diencephalic regions, the similarity of the deficit that occurs after either type of lesion suggests that both of these regions are part of a single brain system that is important for declarative memory.

Although amnesic patients exhibit severe declarative memory deficits, it has become clear over the last 20 years that these patients are capable of learning some kinds of information normally. These nondeclarative learning abilities include perceptual and motor skill learning, and priming, in which the prior presentation of a stimulus allows the stimulus to be processed more fluently and accurately when it is presented a second time (*18*). One characteristic of nondeclarative learning is that subjects are not typically aware of what is being learned. For example, a subject might improve at performing a skill, like a tennis stroke, across days but yet might not be aware of exactly how his or her motor behavior is changing to lead to the improvement. The fact that amnesic patients exhibit normal nondeclarative learning has provided some of the strongest evidence for the idea that memory is not unitary but rather is composed of separate systems.

The dissociation between declarative and nondeclarative memory observed in amnesic patients could potentially shed light on the problem of how category knowledge is organized in the brain. Because amnesic patients exhibit impaired declarative memory for the specific examples that were presented during training, amnesic patients should exhibit impaired category learning if indeed categorization is based on memory for specific instances. However, if category-level knowledge is acquired independently of item knowledge, it raises the possibility that amnesic patients could exhibit normal category learning.

I. ARTIFITIAL GRAMMAR LEARNING

Category learning in amnesic patients has been studied most extensively using the artificial grammar learning paradigm, an example of a rule-based category learning task. In this task, subjects are presented with a series of letter strings that are formed according to the rules of a finite state rule system, or artificial grammar. After viewing a series of such "grammatical" letter strings, subjects are told for the first time that the letter strings they saw were all formed according to a set of complex rules, and that their task will be to decide for a new set of letter strings whether or not each one is formed according to these rules. Although subjects typically cannot describe much about the grammatical rules, they are able to classify new letter strings at a level that is significantly above chance (*15*). Thus, subjects appear to acquire some of the regularities embodied by the grammatical rules. The fact that subjects are not able to articulate what they have learned about the artificial grammar makes this kind of learning similar to other kinds of nondeclarative learning, suggesting that amnesic patients could acquire this information normally. In fact, amnesic patients exposed to grammatical letter strings are able to categorize new letter strings as well as normal subjects can (*4*, *6*). Thus it appears that category learning based on the rules of an artificial grammar does not depend on the brain structures damaged in amnesia. Despite the good performance of amnesic patients on the grammatical classification task, they were nevertheless severely impaired at recognizing the particular letter strings that were presented during training. Thus, a dissociation was obtained between declarative memory for training items used in the recognition task and category-level knowledge that was used in the classification task. These data suggest that in the artificial grammar learning task, different brain systems are involved in the two kinds of acquisition.

The normal performance of amnesic patients in the artificial grammar learning task demonstrates that classification learning does not depend on declarative memory. However, these data do not reveal the nature of the implicitly learned information that gives rise to categorization ability. Three possibilities have been discussed for the form this information might take. One possibility is that subjects are implicitly learning a portion of the actual abstract rules of the grammar (*14*). A second possibility is that subjects are storing whole letter strings in implicit memory during training and using comparisons between the test and training letter strings to make grammaticality judgments (*22*). Test letter strings that are similar to training strings would be endorsed as grammatical more readily than non-similar training strings. A third possibility is that subjects are implicitly learning information that is summarized across the training exemplars, such as the frequency of letter bigrams and trigrams (chunks) that occurred during training. Test items that contain frequently repeated chunks would have high "chunk strength" and would be endorsed as grammatical more readily than items with low chunk strength (*6*, *19*).

It has been difficult to examine the contribution of these factors separately because they are typically confounded with each other. That is, test items that are formed according to the grammatical rules are also likely to be similar to training items, and they are also more likely to have higher chunk strength than are nongrammatical items. There have been a few attempts to unconfound these factors to some extent. These studies found that both rule adherence and similarity to specific training items influenced subjects' classification judgments. However, in at least one of these studies, similarity was confounded with chunk strength such that test items that were similar to specific training items had higher chunk strength than test items that were not similar to any whole training items (*22*). Recently, subjects were tested using materials that were constructed so that chunk strength was the same for similar and nonsimilar test items. Using these materials, the effect of similarity disappeared (*6*). Thus, subjects do not seem to be deciding whether test items are grammatical or not by comparing them to training items stored in memory.

The fact that manipulating chunk strength dramatically affected performance raises the possibility that artificial grammar learning is based solely on rather concrete, exemplar-specific information. Perhaps subjects are not learning abstract grammatical rules at all. In order to examine whether adherence to grammatical rules exerts any influence on grammatical judgments, we constructed a new set of materials, this time balancing chunk strength for items that adhered to the grammatical rules and those that did not. In this way, the effects of rule adherence and chunk strength on judgments could be evaluated separately. Using these materials, there were significant effects of both rule adherence and chunk strength on judgments; that is, subjects were more likely to endorse test items that obeyed the grammatical rules than items that did not, even when chunk frequency was held constant. In addition, subjects tended to endorse items with high chunk frequency for both grammatical and nongrammatical test items (*7*). Thus, it appears that both abstract, rule-based knowledge, and more concrete, exemplar-specific knowledge in the form of chunk frequency are learned in this paradigm. Importantly, data from amnesic patients using these balanced materials has the same pattern as the data from control subjects. Thus, both types of information can be learned implicitly.

In addition to the study just described, data from letter-set transfer experiments support the idea that both abstract, rule-based knowledge and exemplar-specific knowledge contribute to grammaticality judgments. In these experiments, subjects are tested with letter strings comprised of letters different than the ones presented during training. If subjects perform as well when different letters are used at study and test as they do when the same letters are used at study and test, it demonstrates that subjects are primarily learning very abstract information about the artificial grammar. If, however, subjects are unable to show much transfer when tested with the new letter set, it would suggest that they are learning primarily exemplar-specific information. What has been found consistently is the intermediate possibility: subjects perform better with a consistent letter set at study and test, although significant transfer performance does occur across letter sets (*7, 23*). Again, this pattern of results has been obtained with both amnesic patients and control subjects, suggesting that both abstract and concrete information in this task can be learned independently of the brain structures damaged in amnesia.

II. PROTOTYPE ABSTRACTION

Artificial grammar learning is an example of a rule-based category learning task. In contrast, membership in most natural categories is determined by resemblance to a central tendency, or prototype, rather than adherence to a set of rules. Items that resemble the prototype are classified most readily as belonging in the category, while items that are further from the prototype are classified less readily as belonging in the category (*17*). For example, people will classify a robin as a bird more readily than they will a penguin or a chicken. We have studied the ability of amnesic patients to learn such naturalistic categories using materials similar to those used by Posner and Keele in the late 1960s to examine prototype abstraction (*13*). In our experiments, subjects view a series of dot patterns which are distortions of a randomly generated dot pattern. During this training phase, subjects are not exposed to this prototype, but only to the distortions of the prototype. At test, subjects are told for the first time that the dot patterns they had just seen belonged in a single category of dot patterns and that their task would be to decide for a new series of dot patterns whether or not each one belonged in the same category as the dot patterns that were presented earlier. The test consisted of 84 dot patterns: 4 examples of the category prototype, 20 low distortions of the prototype, 20 new high distortions of the prototype, and 40 random dot patterns that served as distractors. Both amnesic patients and control subjects exhibited evidence of prototype abstraction in this task. Both groups endorsed the prototype as being in the category with the greatest frequency, followed by the low distortions of the prototype, then the high distortions of the prototype, with the random dot patterns endorsed the least. The two groups did not differ in terms of classification accuracy (*5*).

Despite the normal performance of amnesic patients on the classification task, they were severely impaired relative to the control subjects on a recognition task for dot patterns. In this task, 5 dot patterns were presented 8 times each during the study phase, and at test these 5 patterns were presented along with 5 distractor patterns. As in the case of artificial grammar learning, there was a dissociation between the performance of the amnesic patients on the classification task and the recognition task. However, one could argue that rather than different memory systems being involved in the task, it could be that some other factor that is different between the two tasks might have differentially affected the amnesic patients. For example, the amount of repetition was different in the two study phases. In the classification task, 40 examples of the category were presented. However, in the recognition task, each item to be remembered was presented only 8 times. Thus, the amnesic patients might have performed more poorly than the control subjects on the recognition task because it involved less repetition than the classification task. To address this issue, we designed another classification task, using different materials, in which only 4 training items were presented. With these materials, amnesic patients again exhibited normal classification performance, suggesting that the amount of training was not the cause of the dissociation between classification and recognition performance seen in the amnesic patients.

There was another possibility we needed to consider. In the recognition task, there were 5 different dot patterns to be remembered. In contrast, in the classification task, subjects really only had to remember one thing, that is, the abstracted prototype. Perhaps this difference in memory load between the two tasks led to the dissociation found in the amnesic patients. In order to address this issue, we constructed a new recognition task. In this test, subjects were shown one dot pattern 4 times, and then tested using this target pattern and a distractor pattern. In this recognition test, the memory load was the same (1 item) as for the classification test. As we predicted, the amnesic patients were still impaired on this recognition test. So it appears that even when recognition and classification tasks are completely parallel, amnesic patients exhibit a striking dissociation in performance between the two tasks. These results provide strong evidence for the idea that category-level knowledge, used for classification, and specific item knowledge, used for recognition, are acquired by separate brain systems.

III. PROBABILISTIC CLASSIFICATION LEARNING

In a third set of experiments, we have examined the ability of amnesic patients to acquire probabilistic category level information (*8*). In these experiments, there were 4 cues which were each probabilistically associated with one of two outcomes. In one of the tasks, the 4 cues were cards with different geometric shapes printed on them (circles, diamonds, squares, or triangles), and the two outcomes were either sunshine or rain. In this task, the subject pretended to be a weather forecaster, and he or she predicted whether rain or sun would occur on each trial based on the combination of cues that appeared. We reasoned that because of the probabilistic nature of the cue-outcome associations in this task, information about a single trial is not nearly as relevant as information abstracted across many trials. Declarative memory is specialized for the rapid acquisition of single-trial information, which is exactly what is needed to perform a recognition task (*21*). In contrast, the kind of abstraction across trials in the probabilistic classification task is more like the kind of abstraction across exemplars that is required to learn category level knowledge. Thus, like the other category learning tasks described earlier, probabilistic category learning may also occur independently of declarative memory.

In addition, the probabilistic classification task also resembles habit learning tasks studied in experimental animals. In these tasks, associations between two stimuli are learned gradually across trials, and learning does not depend on the hippocampus or associated brain regions (*9*, *12*). The amnesic patients did in fact perform normally on each of the 3 probabilistic classification tasks. Both the control subjects and the amnesic patients began each task at about the level of chance, and then improved performance across 50 training trials by nearly 20%. Despite the normal performance of the amnesic patients on the classification task, they nevertheless exhibited impaired performance on declarative memory questionnaires that were administered after each task which assessed subjects' knowledge of facts about the training episode. So again, we obtained a dissociation between classification ability and declarative knowledge. It is important to note that with extended training, control subjects are able to outperform amnesic patients on the classification task, suggesting that eventually normal subjects can acquire declarative knowledge of the cue-outcome associations. Nevertheless, early in training learning appears to be nondeclarative.

The results from the 3 category learning paradigms described in this chapter demonstrate that category learning does not necessarily depend on the brain structures damaged in amnesia, including the hippocampus. Under some conditions, declarative memory could contribute to categorization, such as a situation in which there were only a few highly memorable examples of a category presented during training. However, the fact that category learning can occur independently of the declarative memory system indicates that nondeclarative learning is capable of acquiring high-level cognitive information.

SUMMARY

Despite impairments in memory for facts and events (declarative memory), amnesic patients exhibit an intact capacity for skill learning, priming, and simple forms of conditioning (nondeclarative memory). Recently, it appears that nondeclarative memory also includes abilities that are not primarily perceptual or motor. One such ability is category learning. In three different paradigms, category learning can be accomplished normally by amnesic patients. In the first paradigm, amnesic patients and control subjects were shown letter strings generated by a finite-state rule system (artificial grammar) and were later able to classify new letter strings as to whether or not they adhered to the rules of the artificial grammar. The performance of the amnesic patients was normal on the classification task, although it was impaired when they were asked to recognize the letter strings they had seen previously. In the second paradigm, amnesic patients and control subjects were shown a series of dot patterns that were distortions of a randomly-generated dot pattern (the "prototype"). The two groups did not differ in their ability to classify new dot patterns according to whether or not they were in the same category as the dot patterns that had been presented previously. Both groups also demonstrated that they had abstracted the category prototype, in that this pattern was readily endorsed as being in the category. Again, the amnesic patients exhibited impaired memory for individual dot patterns. In the third set of experiments, amnesic patients demonstrated normal performance in tasks in which the associations between the test items and their categories were probabilistic. These results suggest that category knowledge may be learned in parallel with item information and independently of the medial temporal lobe and diencephalic structures damaged in amnesia.

REFERENCES

1 Cohen, N.J. and Squire, L.R. (1980). *Science* **210**, 207–210.
2 Fried, L.F. and Holyoak, K.J. (1984). *J. Exp. Psychol.: Learn., Mem., Cogn.* **10**, 234–257.
3 Hintzman, D. (1986). *Psychol. Rev.* **93**, 411–428.
4 Knowlton, B.J., Ramus, S.J., and Squire, L.R. (1992). *Psychol. Sci.* **3**, 172–179.
5 Knowlton, B.J. and Squire, L.R. (1993). *Science* **262**, 1747–1749.
6 Knowlton, B.J. and Squire, L.R. (1994). *J. Exp. Psychol.: Learn., Mem., Cogn.* **20**, 79–91.
7 Knowlton, B.J. and Squire, L.R. (1995). *J. Exp. Psychol.: Learn., Mem., Cogn.*, in press.
8 Knowlton, B.J., Squire, L.R., and Gluck, M.A. (1994). *Learn. Mem.* **1**, 106–120.
9 Malamut, B.L., Saunders, R.C., and Mishkin, M. (1984). *Behav. Neurosci.* **98**, 759–769.
10 Mayes, A.R. (1988). *Human Organic Memory Disorders*, New York: Oxford University Press.
11 Medin, D.L. and Schaffer, M.M. (1978). *Psychol. Rev.* **85**, 207–238.
12 Packard, M.G., Hirsh, R., and White, N.M. (1989). *J. Neurosci.* **9**, 1465–1472.
13 Posner, M.I. and Keele, S.W. (1968). *J. Exp. Psychol.* **77**, 353–363.
14 Reber, A.S. (1976). *J. Exp. Psychol.: Hum. Learn. Mem.* **2**, 88–94.
15 Reber, A.S. (1989). *J. Exp. Psychol.: Gen.* **3**, 219–235.
16 Reed, S.K. (1972). *Cogn. Psychol.* **3**, 382–407.
17 Rosch, E. (1973). In *Cognitive Development and the Acquisition of Language*, ed. Moore, T.E., pp. 111–144. New York: Academic Press.
18 Schacter, D.L., Chiu, C-Y.P., and Ochsner, K.N. (1993). *Annu. Rev. Neurosci.* **16**,

159–182.

19 Servan-Schreiber, E. and Anderson, J.R. (1990). *J. Exp. Psychol.: Learn., Mem., Cogn.* **16**, 592–608.

20 Squire, L.R. (1987). *Memory and Brain*, New York: Oxford University Press.

21 Squire, L.R. (1992). *Psychol. Rev.* **99**, 195–231.

22 Vokey, J.R. and Brooks, L.R. (1992). *J. Exp. Psychol.: Learn., Mem., Cogn.* **18**, 328–344.

23 Whittlesea, B. and Dorken, M.D. (1993). *J. Exp. Psychol.: Gen.* **122**, 227–248.

9

MEMORY DEFICIT AND AWARENESS

HIROTAKA TANABE

Faculty of Health and Sport Sciences, Osaka University, Toyonaka, Osaka 560, Japan

The noted amnesic patient H.M. with bilateral medial temporal lesions involving the hippocampal region is always referred to as a representative case of complete anterograde amnesia that causes him or her to forget the episodes of daily life as rapidly as they occur. Even this profoundly amnesic patient is said to be to some extent aware of his memory disorder (*12*), probably based on descriptions such as 'his frequent apologies for what he fears may be considered lapses from good manners, such as forgetting the names of persons to whom he has just been introduced' and 'he often volunteers stereotyped descriptions of his own state, by saying that it is "like waking from a dream"' (*11*). There are, however, no direct descriptions in this article (*11*) demonstrating that H.M. is aware of his inability to form new memories. Could it be that patients with a total inability to lay down new memory traces are really aware of their memory deficits?

In the present study, I examined whether or not complete global amnesics could be aware of their inability to record ongoing information, through their speech and behavior during the amnesic episode in 3 cases of transient global amnesia (TGA) with transient dysfunction confined to the bilateral medial temporal lobe including the hippocampal area.

I. CLINICAL FEATURES OF TGA

TGA is a well-recognized clinical entity characterized as follows (*7, 13*). Patients suffer sudden anterograde and retrograde memory loss without other neurological signs or symptoms. They retain personal identity and are able to carry on normal activities. However, they are unable to acquire any new information, namely, they show complete anterograde amnesia. In addition, their inner world goes back to a certain point of the past due to retrograde amnesia, while the outer world including their appearance is in real time. As a result, various things incomprehensible to them develop in their immediate environment and situation. For example, one patient was astonished at a scar from a burn on her hand she had gotten 5 months before and asked her husband about it. Patients ask questions such as "What day is today?" and "Why am I here?" to try to grasp things incomprehensible to them or orient themselves into that particular situation. However, they cannot retain the answers due to the complete anterograde amnesia, so that they repeat the same questions despite repeated explanations. The retrograde memory loss, extending back for days, months or even years before the attack, gradually or quickly shrinks as the patient recovers. Not only does a brief period of permanent retrograde amnesia persists, but memory for events during the acute phase is permanently impaired. The amnesic episodes generally last several hours.

Thus, a TGA episode gives a rare chance to examine 'awareness' of memory deficit in the phase of different degrees of amnesia in a single subject, and it is considered to be the best material with which to consider the current issue.

II. SUBJECTS

Case S.T. is a 63-year-old, right-handed housewife. One morning she became suddenly perplexed and asked her son the same questions over and over despite repeated explanations. For instance, she asked him where her husband was, evidently not remembering that he had been hospitalized with a fracture of his foot one month before. Her son, who is a medical doctor, took her immediately to our hospital by car. On the way, she repeatedly demanded to know where they were going, despite the fact that she was given the same reply each time. She showed complete inability to form any new memories; she could readily remember 3 object names immediately after they were hidden. However, she could recall none of the objects after distraction such as 5 seriated subtractions of the digit "7" from 100 or after a delay of a couple of minutes. Moreover, she could not remember at all that she had seen them. Her immediate or short-term memory was preserved; digit spans were 7 forward and 5 backward. Her behavior also reflected her dramatic memory loss. At one point, the examiner left the room and returned 5 min later. She behaved as if she had never seen the doctor before. Her retrograde amnesia was considered to cover at least 10 months with fragments of memory during the acute phase.

This case gave me my first experience with this striking amnesic episode of TGA and called my attention to the question whether or not TGA patients are really aware of their memory deficits themselves. In the article on this case (*13*), we wrote as follows; 'During the TGA, our patient appeared perplexed and asked a series of questions repeatedly as if to try to orient herself, but she did not definitely realize her inability to acquire new information'.

Case A.S. is a 57-year-old, right-handed neuropsychiatrist with experience in lecturing on TGA. He had occasionally complained of a migraine for the preceding 3 years. One morning he complained of a headache at his office. Before noon, his secretary wondered why he did not leave his office because he was due to meet his colleagues at noon. When she asked him about the appointment, he had forgotten it and repeatedly asked her the date despite being given the same answer. On the way to our hospital by car, he repeatedly asked where he was going and why he must go to hospital, despite always being given the same reply, like case 1. When he arrived at our hospital, he was still in the acute phase. His digit spans were 8 forward and 7 backward. He shortly became able to retain fragments of some events, so that he had gotten out of the acute phase and entered the recovery stage. His retrograde amnesia was thought to cover 6 weeks at that point and one week 30 min later.

During the acute stage where he not only could recall none of 3 object names after distraction or after a few minutes, but also could not even remember the fact that he had seen them, he never admitted that he suffered from TGA, although he said to us that he knew TGA well and lectured on it to medical students. In the recovery stage where he became able to remember or choose from multiple choice one of 3 objects after distraction or a delay of a few minutes, he admitted at last that he had some problems with his memory.

The examination of the second case demonstrated the following. While he had a complete anterograde amnesia with some retrograde amnesia, he indeed admitted to feeling strange or disoriented in time, but he did not identify the problems as a memory disorder.

Case M.Y. is a 63-year-old, right-handed white-collar worker of a pharmaceutical company. For a few days before the attack of TGA, he had been fatigued due to overwork. One day just after eating his dessert at lunch, his actions suddenly became strange. After he was brought to our hospital, he still repeated the same questions: "Did I do anything wrong ?" and "Why am I here ?," even after receiving adequate replies. At that point, retrograde amnesia covered a few years and he could not remember even his first son's marriage which had been held in Hawaii one year earlier. Like the former cases, his short-term memory was preserved; digit spans were 8 forward and 4 backward.

In this third case, I asked the patient more direct questions of whether or not he was aware that he was suffering from a failure to remember events while at the same time exhibiting a complete inability to store new memories.

M.Y.'s awareness of the nature of his memory deficit is clearly revealed in the following conversation with the author (A=author; P=patient).

A: You have a problem with your memory.

P: Really ?

A: You're aware that you forget things ?

P: No, but I can't quite remember why I'm here now.

A: How about now ? Do you have trouble remembering things ?

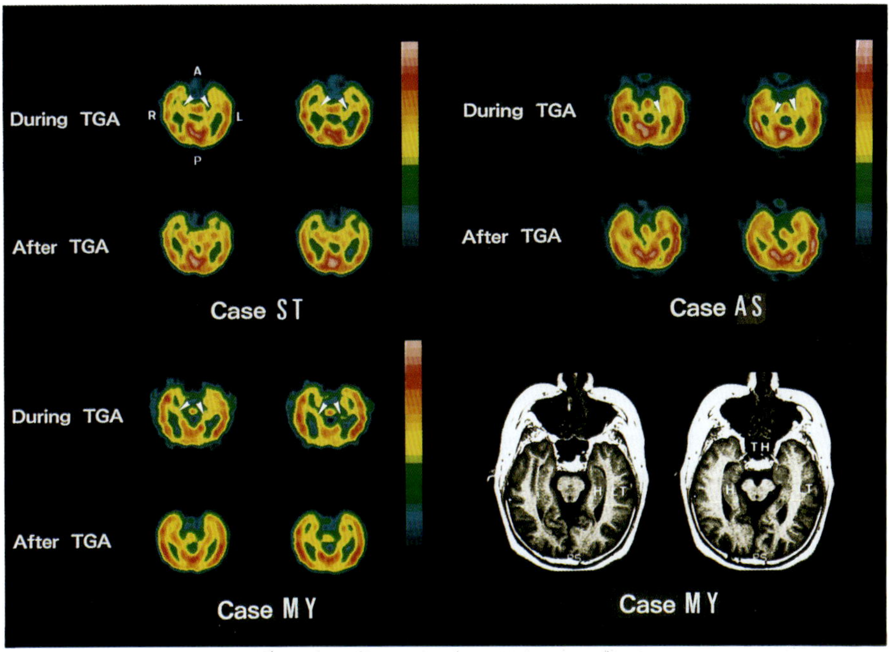

Fig. 1. Hippocampal SPECT images during and after the episode of TGA, using N-isopropyl-(I-123)-*p*-iodoamphetamine (IMP) in case S.T. and Tc-99m hexamethyl-propyleneamine oxime (HM-PAO) in cases A.S. and M.Y. These special images are taken parallel to the longitudinal axis of the hippocampus, so that they enable the evaluation of total blood flow of the hippocampal area in one or two slices (*9, 14, 16*). During TGA, SPECT scans were performed 6 or 7 hr after the onset. After TGA, SPECT scans were administered one week later in case S.T., one week later in case A.S., and the next day in case M.Y. In each case, the upper row of images was taken during the episode and those on the lower row after the episode. In each case, the left and right images are those involving the lower and upper hippocampal region, respectively. Hippocampal MR (magnetic resonance) images of case M.Y., approximately corresponding to the level of the hippocampal slices of SPECT, are shown at the bottom right for anatomical identification of hippocampal SPECT images. In the SPECT and MR images, the left side of the slices corresponds to the right side of the brain. Pink indicates the highest level of cerebral blood flow and blue the lower level. White arrowheads indicate areas with reduced blood flow. Compared with after-TGA images, those taken during TGA demonstrate a definite hypoperfusion in the medial portion of the temporal lobe bilaterally in all the cases. Blood flow in other areas including the frontal cortex and thalamus was preserved. A, anterior; P, posterior; R, right; L, left; H, hippocampus; TH, temporal horn; LT, lateral aspect of the temporal lobe; M, midbrain.

P: Oh, I haven't had much trouble with my memory before.

A: You don't have any problems now?

P: No.

Thus, it is demonstrated that in the acute phase of TGA, patients were aware of particular situations incomprehensible to them resulting from their memory deficits, but unaware of their memory deficits themselves, and that in the recovery phase they at last became aware of their memory problems.

In all 3 cases, single photon emission computed tomography (SPECT) scans taken during the TGA episode revealed hypoperfusion confined to the bilateral medial aspect of the temporal lobe involving the hippocampal region, while those taken after TGA showed no abnormalities (Fig. 1) (*9, 13, 14*).

III. CONCLUSION

Concerning a TGA patient's insight or judgement with respect to his/her amnesic condition, some authors (*5, 12*) argue that he/she is aware of his/her memory loss or profound anterograde amnesia, while other authors (*1, 10*) contradict this argument. For instance, Kushner and Hauser (*10*) and Caplan (*1*) gave the following descriptions, respectively: "Nine patients exhibited insight by observing that something was amiss, but no patient identified the problem as a memory disorder;" and "Some patients admit to feeling strange, disoriented, frightened or somehow ill, though it was rare for them to identify the problem as one involving poor memory".

This discrepancy in opinion seems to be partly due to the phase of TGA during which these doctors examined the patients, that is, whether in the acute or recovery phase, and partly to the difference in their judgements on 'awareness of deficit.' For the latter point, the following should be mentioned. Such complaints of TGA patients described in the literature as: "I'm having a stroke" (*3*), "I'm losing my mind" (*3*), "Am I going mad? I can't remember anything" (*2*), and "There's something wrong with me" (*4*), do not directly imply that they are aware of their memory problems, even though they do indicate they are aware of an unusual situation.

Our investigations into the 3 cases of TGA revealed that during the acute phase of the condition, patients were aware of a particular situation incomprehensible to them resulting from their memory deficit, but unaware of the memory deficit itself, and that in the recovery phase, that is, only after they became able to encode some memories, did they at last become aware of this deficit. This demonstrates that subjects with complete anterograde amnesia cannot be aware of their total inability to acquire new memories.

In other words, awareness of memory deficit is impaired in patients with complete global amnesia in the context that their consciousness is preserved, if consciousness is defined as the awareness of the environment and the self (*6*) or a mechanism that combines attention with short-term memory (Crick and Koch, *8*) (*15*). Their behavior during the TGA episode in which they try to clarify the development and components of that particular situation or try to orient themselves into that situation reflects that they are aware of 'the environment and the self.' In addition, they retain their personal identity and have normal short-term memory.

SUMMARY

It is said that even a dense amnesic patient, such as the noted patient H.M. with bilateral medial temporal lobe resection involving the hippocampal region but without frontal lobe dysfunction, is to some extent aware of his memory disorder. Are patients with complete inability to form new memories really aware of their memory deficit?

I examined whether or not complete global amnesics could be aware of their inability to record ongoing information, based on inquiries taken during the amnesic episode in 3 cases of TGA with transient dysfunction restricted to the bilateral medial temporal lobe. It was of interest that one of them was a neuropsychiatrist with experience lecturing on TGA.

During the acute phase of TGA, patients were aware of a particular situation incomprehensible to them resulting from their memory deficits, but were unaware of the memory deficits themselves. In the recovery phase, they became aware of their memory problems.

Thus, it is demonstrated that unawareness of memory deficit exists in the context of clear consciousness, if consciousness is defined as the awareness of the environment and the self.

REFERENCES

1 Caplan, L.R. (1985). In *Handbook of Clinical Neurology*, ed. Frederiks, J.A.M., vol. 45, pp. 205–218. Amsterdam: Elsevier.
2 Evans, J.H. (1968). *Brain* **89**, 539–548.
3 Fisher, C.M. and Adams, R.D. (1964). *Acta Neurol. Scand. (Suppl.)* **9**, 1–83.
4 Fisher, C.M. (1982). *Arch. Neurol.* **39**, 605–608.
5 Fisher, C.M. (1989). *Neurology* **39**, 127–132.
6 Frederiks, J.A.M. (1969). In *Handbook of Clinical Neurology*, vol. 3, ed. Vinken, P.J. and Bruyn, G.W., pp. 48–61. Amsterdam: Elsevier.
7 Frederiks, J.A.M. (1993). *Clin. Neurol. Neurosurg.* **95**, 265–283.
8 Horgan, J. (1994). *Sci. Am.* (July), 72–78.
9 Kazui, H., Tanabe, H., Ikeda, M., Nakagawa, Y., Shiraishi, J., and Hashikawa, K. *Behav. Neurol.*, in press (1995).
10 Kushner, M.J. and Hauser, W.A. (1985). *Ann. Neurol.* **18**, 684–691.
11 Milner, B., Corkin, S., and Teuber, H.L. (1968). *Neuropsychologia* **6**, 215–234.
12 Schacter, D.L. (1991). In *Awareness of Deficit after Brain Injury; Clinical and Theoretical Issues*, ed. Prigatano, G.P. and Schacter, D.L., pp. 127–151. Oxford: Oxford University Press.
13 Tanabe, H., Hashikawa, K., Nakagawa, Y., Ikeda, M., Yamamoto, H., Harada, K., Tsumoto, T., Nishimura, T., Shiraishi, J., and Kimura, K. (1991). *Acta Neurol. Scand.* **84**, 22–27 and 463.
14 Tanabe, H., Ikeda, M., and Hashikawa, K. (1994). *Shinkei Kenkyu no Shinpo* **38**, 161–172 (in Japanese).
15 Tanabe, H. (1994). Proceedings of International Institute for Advanced Studies Symposium "Cognition, Computation and Consciousness," Aug. 31–Sept. 3, 1994, Kyoto, Japan.
16 Tanabe, H., Kazui, H., Ikeda, M., Hashimoto, M., Yamada, N., and Eguchi, Y. (1994). *Neuropathology* **14**, 105–114.

10

BEYOND THE VISUAL SYSTEM: MULTIPLE MEMORY SYSTEMS IN THE PREFRONTAL CORTEX

SÉAMAS P. Ó SCALAIDHE AND PATRICIA GOLDMAN-RAKIC

Section of Neurobiology, Yale University School of Medicine, New Haven, CT 06510, U.S.A.

The issue of localization of behavioral functions within the nervous system has long been central to the field of neuroscience and is a primary question for one who seeks a materialistic understanding of "Emotion, Memory, and Behavior". It is humbling to think that this question is only now being answered in the higher levels of the visual system and is at the frontier of knowledge in areas, such as prefrontal cortex, that mediate higher cognitive functions.

As is often the case, our conceptualization of the brain is determined as much by the techniques that we use to study it as they are by the brain itself. In the case of cortical organization we have witnessed an evolution from the nineteenth century "cortical organ" concept of phrenology supported by conjecture based on skull shape and personality traits, through anti-localization supported by the effects of lesions on crude measurements of behavior in the middle of the twentieth century, to finally arrive at the current notion of parallel systems even within a single sensory modality, relying largely on evidence from modern anatomical techniques, single unit recording under controlled conditions, and the effects of experimental lesions on specific behavior.

A recent review of the visual system (*10*) listed 32 visual areas and 187 connections between them. Lending some order to this plethora of areas and connections is the organization of areas into a hierarchy based on their pattern of feedforward and feedback connections. Furthermore, Ungerleider and Mishkin (*39*) have proposed that the extrastriate visual areas can by divided into two processing streams: a dorsal stream for the analysis of spatial information terminating in posterior parietal cortex and a ventral stream for the analysis of object information terminating in inferior temporal cortex, or IT. A large number of studies have confirmed and extended this hypothesis (*1, 21, 26, 42*). The synthesis of these ideas about visual organization—that the cortex consists of multiple processing hierarchies based on the sensory submodality that is being processed—is integral to an understanding of contemporary work on visual cortex.

Conspicuous in its absence from most contemporary work on the visual system is an account of the outputs of the visual system, both to nonvisual cortex and to subcortical structures. This review will begin to examine how the visual cortex interacts with the prefrontal cortex and how connections with subcortical structures can shed light on the function of these cortical areas, concentrating on the eye movement system.

I. SPECIALIZATION OF FUNCTION IN THE TEMPORAL LOBE

Bearing in mind the putative dichotomy between object and spatial vision, Ó Scalaidhe *et al.* (*29*) set out to study a relatively unexplored area; the superior temporal polysensory area, or STP, which largely but not exclusively has properties of an area involved in visuospatial functions. Figure 1 shows a lateral view of the monkey brain with the superior temporal sulcus opened up to show the location of STP. STP is strongly connected to areas involved in visuospatial functions such as the parietal cortex (*1, 7, 17, 25–27, 32*), the medial superior temporal visual area (*4*) and the frontal eye fields (FEF) (*2, 16, 18, 33, 36*). It also has a weaker connection with IT cortex. Consistent with its anatomical connections, cells in STP largely resemble those of the parietal cortex, responding selectively to complex types of visual motion, having eye movement related activity, and having very

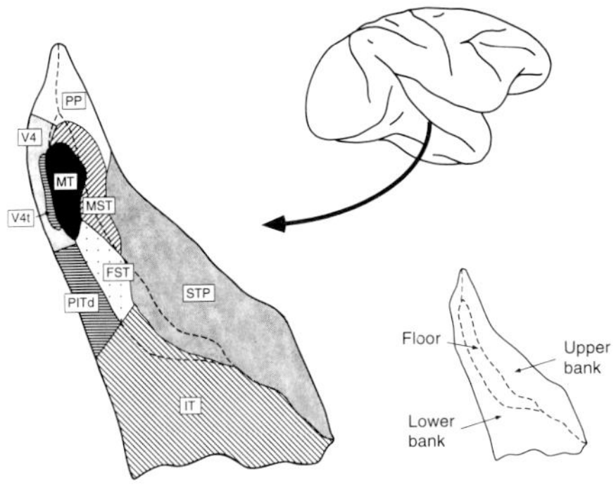

Fig. 1. Opened, "flattened" representation of the superior temporal sulcus illustrating the location of visual areas contained within it, based on Bruce *et al.* (*6*); Felleman and Van Essen (*10*); and Distler *et al.* (*9*). FST, fundus of the superior temporal sulcus visual area; PITd, dorsal portion of the posterior inferior temporal cortex; PP, posterior parietal cortex; V4, visual area V4; V4t, transitional zone of V4.

large receptive fields. Also like the parietal cortex, cells in STP tend not to be selective for stimulus shape, color, orientation, or texture (*6*). One notable exception to this are the small population of cells in the superior temporal sulcus that are selectively responsive to pictures of faces.

To directly test the notion that STP mediates visuospatial functions Ó Scalaidhe *et al.* (*29*) examined the effects of STP lesions on eye movements. The eye movement system is well characterized at both the sensory input and motor output; furthermore, studying eye movements directly obviates the difficulty of making conclusions about visual impairments in the absence of eye position data. Monkeys were trained to visually fixate and to make saccadic and smooth pursuit eye movements, STP was removed, and the monkeys were retested. A group of monkeys with IT lesions served as a control for damage incidental to the STP lesions.

Figure 2A shows the mean saccade latency before and after STP lesions for all of the cases. There was an increase in saccade latency to targets contralateral to the lesion, especially to the more peripheral target locations. Saccade latencies to targets ipsilateral to the lesion and to targets above and below the fixation point were unimpaired. The saccade latency subsequently decreased over the course of four

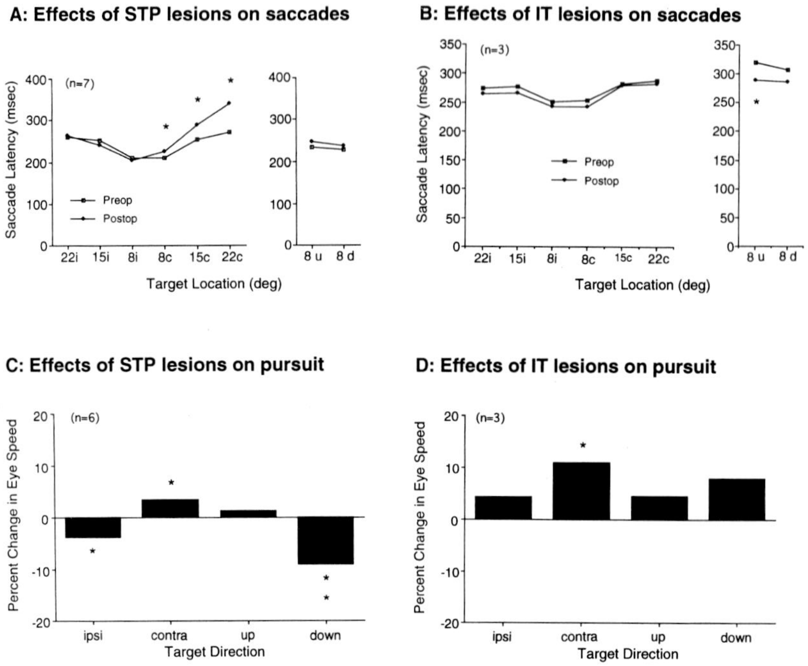

Fig. 2. Effects of STP and IT lesions on eye movements. Mean preoperative and postoperative (first week of testing) saccade latencies for all (A) STP lesions and (B) IT lesions (i=ipsilateral, c=contralateral, u=up, and d=down) and mean percent change in eye speed after all STP lesions (C) and IT lesions (D) to all target speeds. The asterisks denote statistically significant differences (*$p<.05$, **$p<.01$; two tailed *t*-test for paired observations). Asterisks are above the data points for increases and are below the data points for decreases.

weeks of testing until the animals were again performing at preoperative levels. There were no discernible effects on the accuracy of saccadic eye movements. Lynch and McLaren (*23*) observed a mean increase in saccade latency of 37.5 msec (117% of preoperative saccade latency) to targets 24° from the fixation point after lesions of the posterior parietal cortex. This is similar to the effects of STP lesions, which produced an increase in saccade latency of 67.8 msec (125% of preoperative saccade latencies) to targets 22° from the fixation point contralateral to the lesion. In contrast, lesions of IT cortex in the lower bank of the superior temporal sulcus did not affect saccadic eye movements (see Fig. 2B), in fact, there was a tendency for saccade latency to decrease slightly in these monkeys, probably as a result of practice. Neither STP lesions nor IT lesions affected the monkeys' ability to fixate.

Lesions of STP also impaired the monkeys' ability to make smooth pursuit eye movements (see Fig. 2C). Pursuit speed was most impaired to downward moving targets, less impaired to ipsilateral moving targets, unimpaired to upward moving targets, and was increased to contralateral moving targets. These effects were larger to faster moving targets and after bilateral lesions. Like the saccade task results these effects are similar to those seen after parietal lesions on pursuit, which impaired pursuit to vertically moving targets after bilateral lesions (*22*). Lesions of IT did not impair smooth pursuit, instead there was a tendency for eye speed to increase slightly (see Fig. 2D).

While much remains to be learned about STP, our results suggest that it plays a role in visuospatial behavior similar to that of posterior parietal cortex and tentatively place it in the dorsal processing stream.

II. SPECIALIZATION OF FUNCTION IN THE PREFRONTAL CORTEX

1. Eye Movements

Hemidecortication (*37*), combined lesions of posterior parietal cortex and the FEF (*23*), or combined lesions of the FEF and the superior colliculus (*31*) produce deficits that are substantially larger than those that follow individual extrastriate lesions. To further explore the role of STP in visuomotor behavior lesions of the FEF were added to some of the animals with bilateral lesions of STP or IT (*34*). In addition to beginning to determine which areas are involved in recovery after loss of STP, this addressed the question of whether the frontal eye fields and the posterior visual areas are merely redundant or if they play heretofore unrealized specific roles in eye movements.

Not only did addition of FEF lesions to STP lesions produce eye movement deficits larger than those seen after either lesion alone, but they produced effects that are qualitatively different than those seen after lesions of STP. After all FEF lesions, either in combination with STP or IT lesions, there was a profound impairment in fixating the central fixation point. It was discovered that by turning a dim light on in the testing room this impairment could be completely eliminated. Figure 3 shows data from a single monkey over the first, second, and third days of postoperative testing. The same effect was observed after a combined FEF and IT lesion, suggesting that it is loss of the FEF which causes the impairment. This impairment shows that rather than being merely redundant with posterior visuomotor areas, the FEF is specialized for certain kinds of eye movements. Given that all four monkeys with FEF lesions had identical severe impairments in fixating targets in the dark while being able to fixate normally in

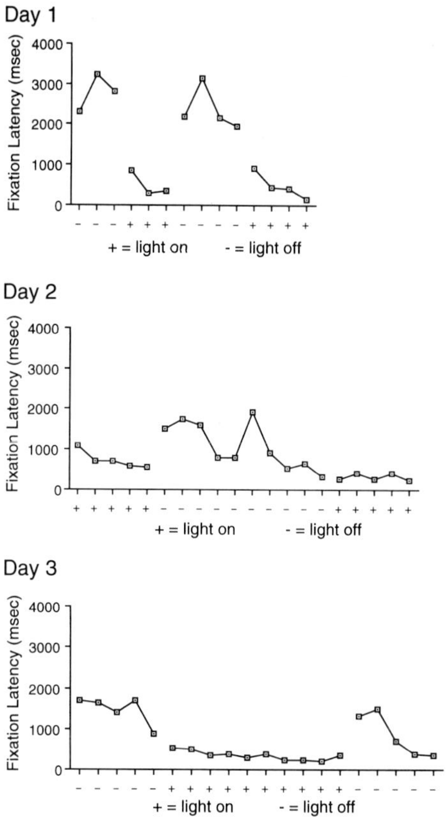

Fig. 3 Fixation latency in darkness and dim illumination for monkey 2 on the: first, second, and third day of testing after a combined STP and FEF lesion. Each data point represents the mean fixation latency from a block of ten trials.

dim illumination we concluded that loss of the FEF impairs the ability to make saccades in the absence of visual landmarks.

Unlike STP lesions alone, there were saccadic accuracy impairments after all FEF lesions. Large neglect-like deficits in making saccadic eye movements, however, were only seen after combined STP and FEF lesions and were similar to those seen by Lynch and McLaren (*23*) after combined posterior parietal cortex and FEF lesions. These results suggest that loss of both the FEF and STP has a synergistic effect on saccades to visual targets even if the deficit is primarily one of saccade latency after loss of STP and saccade accuracy after loss of the FEF. There were also large impairments in making smooth pursuit eye movements after combined STP and FEF lesions, but these were within the range of deficits seen after FEF lesions alone. This suggests a more critical role for the FEF in making smooth pursuit eye movements than STP or the posterior parietal cortex.

2. Spatial Memory

While the visuospatial processing stream continues from STP and posterior parietal cortex into the prefrontal cortex, prefrontal cortex lesions do not cause visual impairments and instead impair the ability to perform working memory tasks (*12*). Accordingly, we have examined the responses of prefrontal neurons on a visuospatial memory task. Previous studies in the Goldman-Rakic laboratory (*11*) had shown that prefrontal neurons were tuned for the direction of a remembered target. It was not determined whether the neurons simply coded polar direction or also reflected the location with relation to distance from the fovea. We recently examined this issue by describing the spatial tuning of cells on an oculomotor delayed response task (*28*). First the monkey fixated a central fixation point, half a second later a spot of light was presented at one of 25 locations throughout the visual field, then the cue was extinguished, the monkey maintained fixation for 3 sec, and finally the fixation point was extinguished, signaling the animal to make a saccade to the location where the cue had been presented. The delay activity was tuned for the distance of the stimulus from the fovea as well as for its direction. Delay activity occurred to contiguous sets of targets ranging in size from a single target to an area larger than a visual quadrant and was usually confined to contralateral targets. By analogy with visual receptive fields we have termed the areas over which these cells show delay related activity "memory fields". There appears to be at least a local organization of delay activity with memory fields in adjacent locations typically occurring on a penetration in which multiple delay cells are encountered. Furthermore, delay cells were concentrated in the FEF and the periarcuate cortex, which receives input from both posterior parietal cortex and STP.

3. Object Memory

To further examine the issue of specialization of function within the prefrontal cortex, we have also been describing the responses of cells throughout the prefrontal cortex to visual stimuli, such as pictures of objects (*40*). In agreement with the data from the oculomotor delayed task, but in contrast with most previous electrophysiological studies of the prefrontal cortex, we found that cells selective for these stimuli are concentrated in the inferior convexity (IFC) anterior to the lower limb of the arcuate sulcus and ventral to the principal sulcus. Testing with this task also revealed a number of face selective cells, all in the IFC. Like IT cortex the face selective neurons in the inferior convexity were intermixed with other cells selective for pictures of objects. In contrast, cortex dorsal to the principal sulcus was nearly devoid of selective responses to these stimuli.

To determine if the specialization of the IFC reflected in the visual responses would also hold for performance on memory tasks we examined cells' responses on a variant of the oculomotor delayed response task described above. This task required the monkeys to make a leftward or rightward saccade based on the identity of a stimulus that they had seen 2.5 sec ago. Figure 4 shows representative responses of one cell to four of the stimuli on this task. This cell responded selectively to one of the faces during the delay period. Note especially that the behavioral response is the same to both of the stimuli requiring a leftward saccade, and as such the selectivity of the delay activity cannot be reflecting motor planning. Similar to the selectivity for pictures of objects and faces, the delay activity related to the memory of pictures was concentrated in the inferior

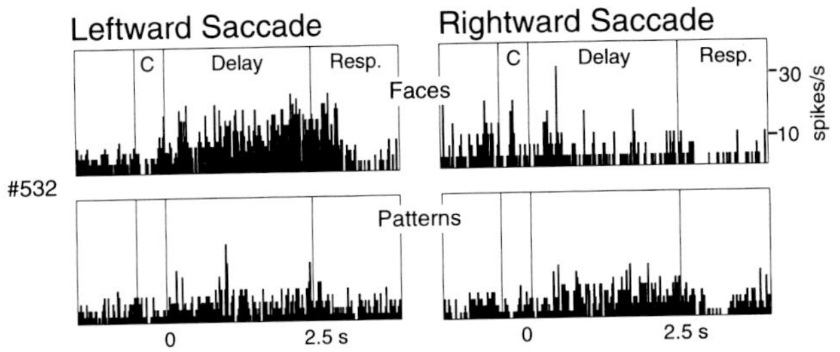

Fig. 4. Responses of inferior convexity neuron #532 with delay-period activity following a specific monkey face in the memory task. Upper panels show differential responses to two faces (requiring left or right responses); lower panels show the lack of response to two patterns requiring left or right responses. The first vertical line indicates the onset of the cue, the second vertical line indicates the offset of the cue and the beginning of the delay, and the third vertical line indicates the offset of the fixation point and end of the delay period.

convexity and was absent in the cortex dorsal to the principal sulcus.

Given that the visual responses that we recorded from the IFC have a close resemblance to those of IT cortex (*14*) we have begun to examine the connections of this area using wheat germ-agglutinin conjugated horseradish peroxidase (HRP) injected into the region from which we recorded visually selective neurons. Consistent with our electrophysiological results we have found (*3*) that the physiologically defined region of the IFC receives input from IT cortex.

III. DISTRIBUTED PARALLEL HIERARCHIES WITH INDEPENDENT OUTPUTS

These results confirm that the multiple streams of processing in the visual system continue from the visual system to the prefrontal cortex (see Fig. 5A). Given the lack of visual discrimination impairments, such as prosopagnosia, after prefrontal cortex damage, it would be inaccurate to think of the prefrontal areas as yet another visual processing stage. Instead, it appears that visual information reaches the prefrontal cortex in a submodality specific fashion and that its use reflects the specialization of the prefrontal cortex for working memory. Furthermore, consideration of subcortical outputs is indicative of the role that the prefrontal areas play in information processing hierarchies that begin in the visual system.

The posterior visual cortex and the prefrontal cortex have

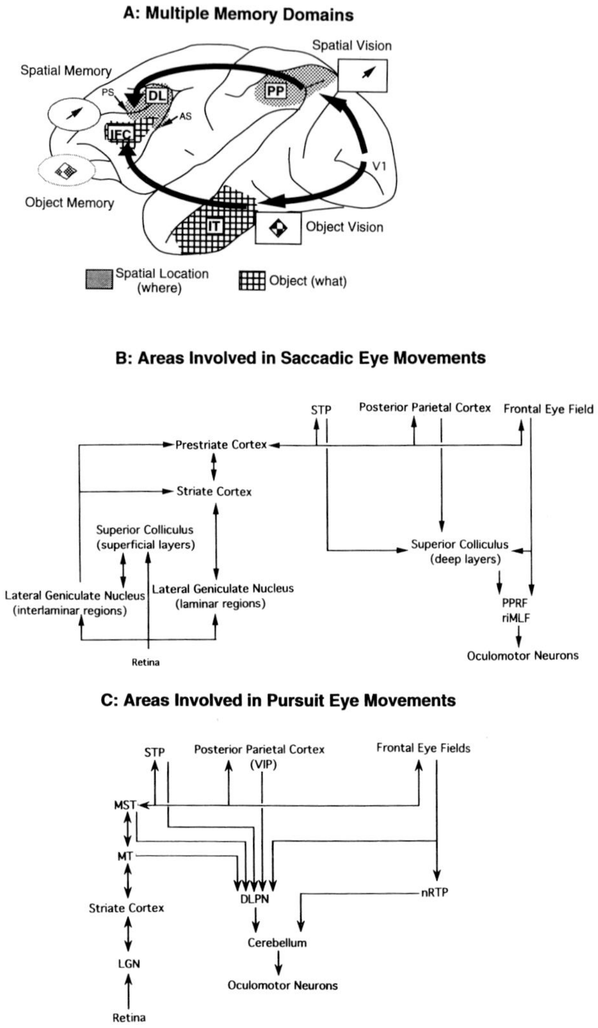

Fig. 5. A: schematic illustration of the primate visual systems spatial and object vision and their connections with prefrontal cortex. The PP cortex is concerned with spatial perception, and IT cortex is involved in object recognition. These regions are connected with the dorsolateral (DL) and IFC prefrontal cortices where memoranda pertaining to spatial location and object identity are encoded in working memory. B: diagram of areas known to be involved in making saccadic eye movements. PPRF, paramedian pontine reticular formation; riMLF, rostral interstitial nucleus of the medial longitudinal fasciculus. C: diagram of areas known to be involved in making smooth pursuit eye movements. LGN, lateral geniculate nucleus.

independent outputs to the motor control areas. This is clearest in the case of saccadic eye movements (see Fig. 5B) where the FEF plays a larger role in fixation in the dark, saccades to remembered locations, and predictive saccades (*5, 8, 19*) than do extrastriate visual areas. This contrasts with extrastriate visual cortex and its projection to the superior colliculus, which appears to be differentially involved in more highly stimulus driven behavior. For example, lesions of the superior colliculus eliminate the ability to make reflex-like express saccades, while FEF lesions do not affect this ability (*30*).

A similar situation may obtain with the pursuit eye movement system (see Fig. 5C) given that FEF lesions impair predictive pursuit (*19*). The pontine outputs of middle temporal visual area (MT) and medial superior temporal visual area (MST) (*38*) also suggest that more stimulus driven smooth eye movements may not necessarily involve the FEF. The projection to the dorsolateral pontine nucleus (DLPN) resembles the projection to the superior colliculus in that most of extrastriate visual cortex projects to this nucleus. Stimulation of the DLPN only affects ongoing smooth pursuit (*24*), while stimulation of the nucleus reticularis tegmenti pontis (nRTP), which receives a direct projection from the FEF (*15, 20, 35*), can produce pursuit from stationary eye positions (*13, 41*).

We suggest that, in general, relatively stimulus driven tasks do not recruit the prefrontal cortex and can be mediated directly by outputs of the visual system. Thus each area within a specialized pathway has a hierarchical relationship to other areas in the pathway, while parallel processing occurs between pathways and within areas. Finally, there are subcortical outputs at a number of stages in each processing hierarchy. This allows the prefrontal areas to be specialized for complex behavior not driven by the immediate environment, such as working memory, while at the same time being an integral part of visual information processing hierarchies.

SUMMARY

1. The extrastriate visual cortex has been divided into multiple processing hierarchies based on visual submodality.
2. On the basis of the effects of lesions on eye movements the

superior temporal polysensory area (STP) appears to be part of the spatial processing hierarchy.
3. The prefrontal cortex is also specialized on the basis of visual submodality, in a manner consistent with its connections to the extrastriate visual cortex.
4. Rather than being yet another stage of visual processing the prefrontal areas are involved in mediating behavior when there is a discontiguity between the relevant stimuli and the behavioral response. By contrast, separate subcortical outputs of the extrastriate visual cortex allow direct control of stimulus driven behavior without obligatory prefrontal involvement.

REFERENCES

1 Baizer, J.S., Ungerleider, L.G., and Desimone, R. (1991). *J. Neurosci.* **11**, 168–190.
2 Barbas, H. and Mesulam, M-M. (1981). *J. Comp. Neurol.* **200**, 407–431.
3 Bates, J.F., Wilson, F.A.W., Ó Scalaidhe, S.P., and Goldman-Rakic, P.S. (1994). *Soc. Neurosci. Abstr.* **20**, 434.10.
4 Boussaoud, D., Ungerleider, L.G., and Desimone, R. (1990). *J. Comp. Neurol.* **296**, 462–495.
5 Bruce, C.J. and Borden, J.A. (1986). *Soc. Neurosci. Abstr.* **12**, 297. 2.
6 Bruce, C.J., Desimone, R., and Gross, C.G. (1981). *J. Neurophysiol.* **46**, 369–384.
7 Cavada, C. and Goldman-Rakic, P.S. (1989). *J. Comp. Neurol.* **287**, 393–421.
8 Deng, S., Goldberg, M.E., Segraves, M.A., Ungerleider, L.G., and Mishkin, M. (1986). In *Adaptive Processes in Visual and Oculomotor Systems*, ed. Keller, E.L. and Zee, D.S., pp. 201–208. London: Pergamon.
9 Distler, C., Boussaoud, D., Desimone, R., and Ungerleider, L.G. (1993). *J. Comp. Neurol.* **334**, 125–150.
10 Felleman, D.J. and Van Essen, D.C. (1991). *Cerebral Cortex* **1**, 1–47.
11 Funahashi, S., Bruce, C.J., and Goldman-Rakic, P.S. (1989). *J. Neurophysiol.* **61**, 331–349.
12 Goldman-Rakic, P.S. (1987). In *Handbook of Physiology—The Nervous System*, ed. Plum, F., vol. V, pp. 373–417. Bethesda, MD: American Physiological Society.
13 Gottlieb, J.P., Bruce, C.J., and MacAvoy, M.G. (1993). *J. Neurophysiol.* **69**, 786–799.
14 Gross, C.G. (1992). *Philos. Trans. R. Soc. Lond. B* **335**, 3–10.
15 Huerta, M.F., Krubitzer, L.A., and Kaas, H.H. (1986). *J. Comp. Neurol.* **253**, 415–439.
16 Jacobson, S. and Trojanowski, J.Q. (1977). *Brain Res.* **132**, 209–233.
17 Jones, E.G. and Powell, T.P.S. (1970). *Brain* **93**, 793–820.

18 Kawamura, K. and Naito, J. (1984). *Neurosci. Res.* **1**, 89–103.
19 Keating, E.G. (1991). *Exp. Brain Res.* **86**, 311–323.
20 Leichnetz, G.R., Smith, D.J., and Spencer, R.F. (1984). *J. Comp. Neurol.* **228**, 388–408.
21 Livingstone, M.S. and Hubel, D.H. (1988). *Science* **240**, 740–749.
22 Lynch, J.C. (1982). In *Functional Basis of Ocular Motility Disorders*, ed. Zee, D.S. and Keller, E.L., pp. 501–510. London: Pergamon.
23 Lynch, J.C. and McLaren, J.W. (1989). *J. Neurophysiol.* **61**, 74–89.
24 May, J.G., Keller, E.L., and Crandall, W.F. (1985). *Soc. Neurosci. Abstr.* **11**, 25.12.
25 Mesulam, M.-M., Van Hoesen, G.W., Pandya, D.N., and Geshwind, N. (1977). *Brain Res.* **136**, 393–414.
26 Morel, A. and Bullier, J. (1990). *Vis. Neurosci.* **4**, 555–578.
27 Neal, J.W., Pearson, R.C.A., and Powell, T.P.S. (1988). *Brain Res.* **438**, 351–356.
28 Ó Scalaidhe, S.P. and Goldman-Rakic, P.S. (1993). *Soc. Neurosci. Abstr.* **19**, 326.6.
29 Ó Scalaidhe, S.P., Albright, T.D., Rodman, H.R., and Gross, C.G. (1995). *J. Neurophysiol.* **73**, 1–20.
30 Schiller, P.H., Sandell, J.H., and Maunsell, J.H.R. (1987). *J. Neurophysiol.* **57**, 1033–1049.
31 Schiller, P.H., True, S.D., and Conway, J.L. (1980). *J. Neurophysiol.* **44**, 1175–1189.
32 Seltzer, B. and Pandya, D.N. (1984). *Exp. Brain Res.* **55**, 301–312.
33 Seltzer, B. and Pandya, D.N. (1989). *J. Comp. Neurol.* **281**, 97–113.
34 Skelly, J.P., Albright, T.D., Rodman, H.R., and Gross, C.G. (1991). *Soc. Neurosci. Abstr.* **17**, 338.13.
35 Stanton, G.B., Goldberg, M.E., and Bruce, C.J. (1988). *J. Comp. Neurol.* **271**, 473–492.
36 Stanton, G.B., Goldberg, M.E., and Bruce, C.J. (1989). *Soc. Neurosci. Abstr.* **15**, 475.7.
37 Tusa, R.J., Zee, D.S., and Herdman, S.J. (1986). *J. Neurophysiol.* **56**, 1590–1625.
38 Tusa, R.J. and Ungerleider, L.G. (1988). *Ann. Neurol.* **23**, 174–183.
39 Ungerleider, L.G. and Mishkin, M. (1982). In *Analysis of Visual Behavior*, ed. Ingle, D.J., Goodale, M.A., and Maunsfield, R.J.W., pp. 549–586. Cambridge, MA: MIT Press.
40 Wilson, F.A.W., Ó Scalaidhe, S.P., and Goldman-Rakic, P.S. (1993). *Science* **260**, 1955–1958.
41 Yamada, T., Suzuki, D.A., Betelak, K.F., and Yee, R.D. (1991). *Soc. Neurosci. Abstr.* **17**, 185.12.
42 Zeki, S.M. and Shipp, S. (1988). *Nature* **335**, 311–317.

11

PLACE-CORRELATES OF MONKEY HIPPOCAMPAL NEURONAL ACTIVITY AND ITS SIGNIFICANCE IN MEMORY

TAKETOSHI ONO, SATOSHI EIFUKU, HISAO NISHIJO, AND RYOI TAMURA

Department of Physiology, Faculty of Medicine, Toyama Medical and Pharmaceutical University, Toyama 930-01, Japan

Hippocampal formation (HF), a structure in the medial temporal lobe which is composed of the hippocampus proper, the dentate gyrus, and the subicular complex (*2*), has been one of the most mysterious structures in the brain, although its behavior has attracted the attention of many cognitive neuroscientists. As we all know, two early studies are distinctive. Scoville and Milner (*11*) first disclosed the role of HF in the intermediate memory of human. Patient H.M. with bilateral resection to the medial temporal lobe including the HF manifested not only anterograde but also retrograde amnesia that was temporally graded. O'Keefe and Dostrovsky (*6*) (also, see the most recent progress in (*14*)) discovered the rat "place-cell" in the HF, which was the first important report of the HF neuronal correlates with behavior of rat. Place-cells specifically fire when the subject is passing through a particular location that is called a "place-field", and we have now attained a consensus that the HF neuronal correlates are related to internal representation of the environment.

The aims of these two studies, however, have been kept separated until today. O'Keefe and Dostrovsky's original study referred to "place" based on ongoing exploratory behavior, whereas Scoville and

Milner referred to a "memory". We need to reconcile this difference to gain a satisfactory understanding of the behavior of the HF. Seeing that hippocampal involvement in memory is clear, we have to establish a new framework in which to link these two extremes.

Zola-Morgan and Squire (*15*) showed clear evidence of a limited-time role of the HF in memory storage; in their prospective study, monkeys with HF damage showed clear anterograde and temporally graded retrograde amnesia in a delayed retention task of concurrent object discrimination. This reproduced the results of Scoville and Milner (*11*) in monkeys in an experimentally controlled way. They offered an explanation of the time-limited involvement of the HF: a kind of organization of information occurred in the HF in the critical months and thereafter such information was transferred to the neocortex. In this review article, based on our previous studies, we try to provide a supplement to this interpretation of the time-limited involvement of the HF; our conceptual framework includes two key points. One is that the HF is crucial for organization of a class of memory, which we should call "declarative" (*12*) or "relational" (*1, 2*), and the other is that such an organization of the memory in the HF is essentially based on the place and context in which a subject is. Here, we show some experimental findings of the monkey HF neuronal correlates to support our framework: place-correlates of the monkey HF neurons and influence of the context on them.

I. PLACE- AND CONTEXT-CORRELATES OF MONKEY HIPPOCAMPAL ACTIVITY

Recently we reported evidence for place-correlates of the monkey HF neurons (*7–9*). In a series of studies, extracellular single unit activity was recorded from three adult monkeys (*Macaca fuscata*), while the monkey maneuvered a cab to various locations as he performed a task (the spatial moving task: Fig. 1). During a trial in the task the monkey was first presented with a 0.5 sec cue tone, then after a 0.7 sec delay, was presented with a 0.5 sec bar light emitting diode (LED) cue. Both cues indicated the appropriate bar that the monkey should press later. A stimulus such as food or an object associated with delivery of a reward (juice) or non-reward was presented after 0.8 sec delay, and if the stimulus was food or an object associated with reward, the monkey could voluntarily press the indicated bar to begin or continue the procurement process. As long as the monkey pressed the correct bar, the cab moved a certain distance (up to 62.5 cm at 5–10 cm/sec), and upon arrival at each step the monkey obtained food or a juice reward. The monkey usually faced in the +Y direction (see Fig. 1), and could see numerous spatial cues in the room.

Increase or decrease of the activity during each intertrial interval (ITI, at least 1 min) in each location was considered to be significant if it exceeded 2.0 standard deviation (SD) from the overall mean of the activity of 16–25 different locations. Task-related neuronal responses were determined by analysis of variance between the pretrial discharge rate during ITI and the rate in each phase of the task with significance level at $p<0.05$.

Of 238 neurons recorded in the monkey HF, 79 (33.2%) had place-related responses; their neuronal activity was higher when the monkey was in some specific location in the experimental field. Task-related activity was evident in 110 (46.3%) of the neurons, in one or more phases of the task: cue signals (cue tone: $n=82$, bar LED: $n=79$); stimulus presentation (food or rewarding/non-rewarding object: $n=104$); bar pressing ($n=59$) and holding (moving $n=94$); reward (food or juice) delivery ($n=26$). Of these 110 neurons, 33 (13.9%) also had place-related responses. Histology after recording revealed place- and task-related neurons distributed throughout the hippocampus proper (CA1, $n=11$; CA3, $n=14$), the dentate gyrus ($n=4$), the subiculum ($n=2$), and the parahippocampal gyrus ($n=2$), and many in the pyramidal cell layer in CA1 and CA3.

Figure 2 shows a typical example of a place-related neuron. The activity of this neuron increased markedly in locations P11-P12 in the field without directional or task related activity changes (A). The neuronal response in P11-P12 was also repeated when alternative routes were used to reach these locations, although different bars were pressed and different elapsed times were needed to arrive at the same locations (not shown). For this neuron, directional stimulus (human movement) from any direction had no obvious effect on neuronal activity when the monkey was at P0 (B), or at P12 (C).

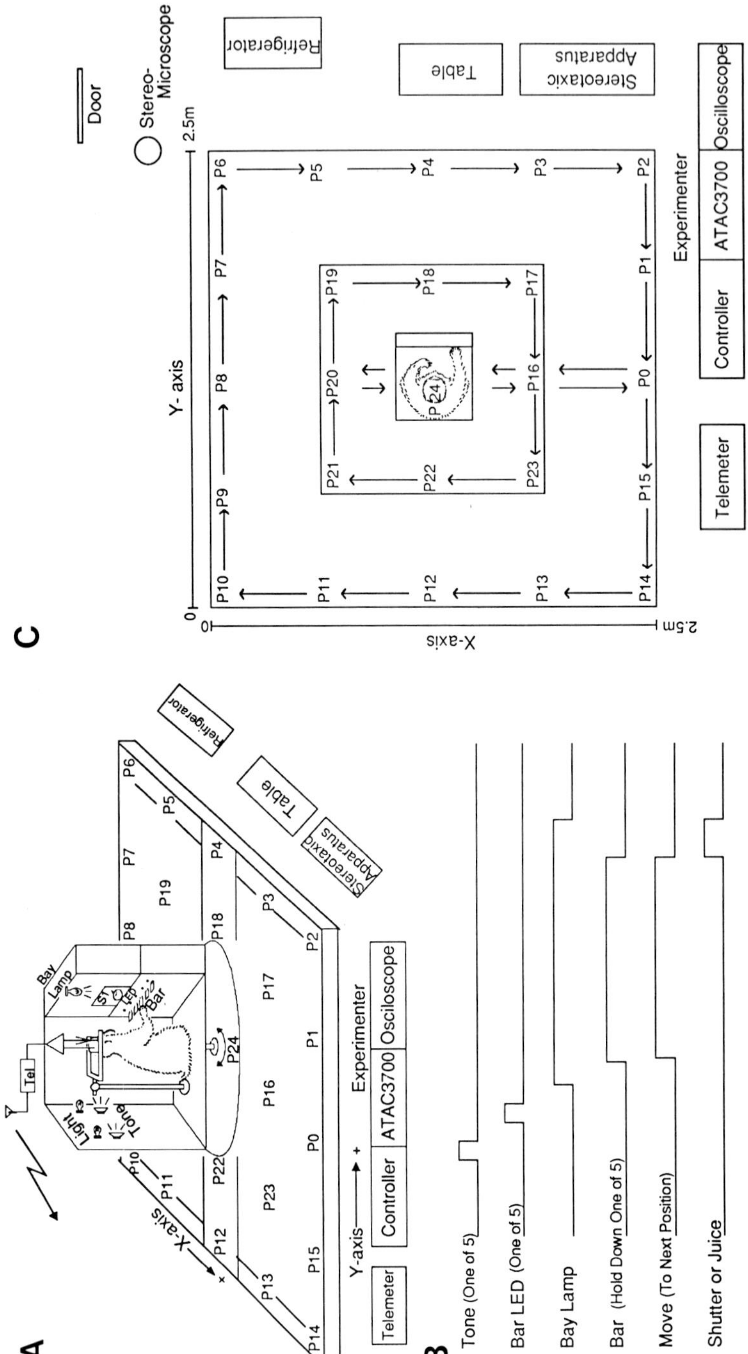
A
Tel
Light
Tone
Bay Lamp
ST
LED
Bar
P24
X-axis
Y-axis
Telemeter
Controller
ATAC3700
Osciloscope
Experimenter
Refrigerator
Table
Stereotaxic Apparatus
P0 P1 P2 P3 P4 P5 P6 P7 P8 P10 P11 P12 P13 P14 P15 P16 P17 P18 P19 P22 P23
B
Tone (One of 5)
Bar LED (One of 5)
Bay Lamp
Bar (Hold Down One of 5)
Move (To Next Position)
Shutter or Juice
Location (X or Y)
Control
Discrim.
Respond
Reward
C
Door
Stereo-Microscope
Refrigerator
Table
Stereotaxic Apparatus
Y- axis
X-axis
2.5m
0
Experimenter
Telemeter
Controller
ATAC3700
Oscilloscope
P0 P1 P2 P3 P4 P5 P6 P7 P8 P9 P10 P11 P12 P13 P14 P15 P16 P17 P18 P19 P20 P21 P22 P23 P24

When the monkey's orientation was changed 90° clockwise at location P12, there was also no significant directional response (D). The overall response intensity increased slightly in this location, since the mean neuronal activity increased slightly (4.5 spikes/sec increase). Therefore, the origin of the stimulus source that induced this place related response could be attributed to something in a fixed location in the experimental room that could not be identified by the experimenter. This neuron also had no significant response to any task cue, *i.e.*, the neuronal activity did not change more than 2.0 SD from that of the pretrial control.

←Fig. 1. Paradigm of a spatial moving apparatus for the monkey, and experimental room cue configuration. A: experimental apparatus. The monkey sat in a chair in the cab (0.7 m long × 0.8 m wide × 0.85 m high). The cab was powered to move within a 2.5 m × 2.5 m space. Front, left and right walls of the cab were half-mirrored. The rear wall, made of steel, was equipped with speakers and a cab light. Five bars and indicator LEDs were on the lower half of the cab front wall. A half mirrored shutter (15 cm × 15 cm) in the upper half of the cab front wall was opened by air pressure to deliver food. A lamp in the food bay illuminated an object in the bay. The cab was on a pivot and could be rotated. The monkey usually faced in the +Y direction in the spatial moving task. The entire apparatus was in a 5 m × 6 m, shielded experimental room. The monkey could see the room environment through the right, front and left half-mirrored walls if the room was illuminated. Locations where the monkey could get a reward and remain until the next trial were numbered from P0 to P24. Each session started from P0 and returned to P0. Directional stimulation was administered after neuronal responsiveness to the spatial moving task was tested. The cab together with the monkey could be rotated to an angle as far as ±90° from the usual orientation, facing in the +Y direction. B: time chart of the spatial moving task. A 0.5 sec cue tone at one of 5 different frequencies started a trial: 5,200 Hz indicated to press the far right bar to move backward, 1,700 Hz indicated to press the right bar to move right, 1,100 Hz indicated to press the middle bar to move forward, 620 Hz indicated to press the left bar to move left, and 210 Hz (usually not used) indicated to press the far left bar to move backward. The tone was followed after a 0.7 sec delay by a 0.5 sec LED light over the corresponding bar. The front bay lamp turned on after 0.8 sec delay to show food, an object associated with juice, or nonfood. If a reward was presented, the monkey pressed the appropriate bar. As long as the monkey held the correct bar down, the cab moved at 5–10 cm/sec toward the goal location 62.5 cm away. At the goal the front shutter opened, and the monkey got food or a drop of juice delivered by a spout in front of its mouth. If the monkey did not hold the bar long enough, or pressed an incorrect bar, the bay lamp was switched off, and the cab stopped until the correct bar was pressed. If the monkey did not press the correct bar within an allowed short period (1.0 sec), that trial was discarded. C: spatial arrangement in the experimental room. One route and the sequence is shown by arrows. Alternative routes and sequences were also used.

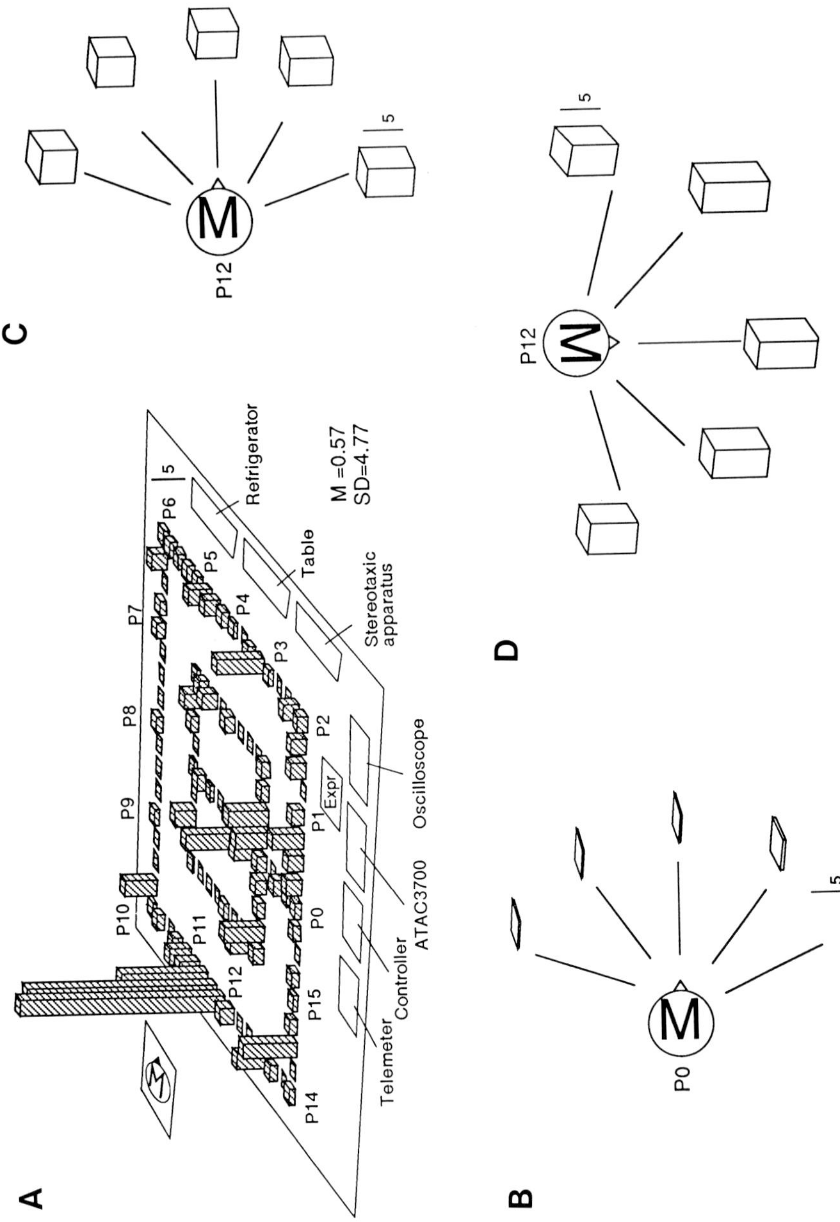
A
B
C
D
P0
P1
P2
P3
P4
P5
P6
P7
P8
P9
P10
P11
P12
P14
P15
Expr
Refrigerator
Table
Stereotaxic apparatus
Oscilloscope
ATAC3700
Controller
Telemeter
M =0.57
SD=4.77
5
M

Fourteen place-related responses were further analyzed under the condition where the experimenter controlled the moving cab without imposing any operant task on the monkey, and compared with those under the condition where the animal solved the task and moved the cab around the room (not shown). The place-related activities of 12 HF neurons turned out to be obscure under the experimenter-controlled condition.

Another characteristic of these place-related neurons is dependence of their sensitivity to external cues on the context in which the monkey is. Figure 3 shows another example of neuronal responses in the spatial moving task in which there were also place related responses (A). This neuron responded to cue signals (both auditory and visual), and to shutter opening for reward, only when the monkey was in locations P11-P7 (Ba,b,c), where place related responses also occurred (A). This neuron did not respond to any task cues in any location in which there was no place related response (Bd). The neuronal activity increased significantly in and around locations P11-P8, and these increases were confirmed by repeating them up to 3 times in the same locations (P11-P8) using the same route. Place related neuron firing and place dependent neuronal responses to task cues by place related neurons required visual signals from outside of the cab. When the monkey performed the spatial moving task with the room light off, the activity increase of the neuron at P11-P8 shown in Fig. 3 did not occur, and there were no responses to task cues in this dark room situation (not shown).

These results imply that the monkey hippocampal neurons can associatively encode multiple cues such as place and event (sight and

←Fig. 2. Place related neuron response with neither directional nor task related responses. A: increased activity of the neuron at locations around P12. The neuronal activity in response to cues did not change ± 2.0 SD from the increased activity in locations P12-P11 in the spatial moving task. Each bar indicates neuronal activity (spikes/sec) in each location. M, overall mean of neuronal activity (spikes/sec). Inset: direction of the monkey during the spatial moving task is indicated. B–D: activity of the neuron during presentation of directional cues. No response to human movement from any direction, when the monkey was at location P0 facing $+Y$ direction (A), when monkey was facing in the $+Y$ direction and located at P12 (C), and when monkey orientation was changed to $+90°$ at location P12 (D). Vertical bars: spikes/sec.

ingestion of an apple), and that place may work as the context to evoke response to an event. Recently, we compared the monkey HF neuronal responses between performance of a place-dependent conditional association (PCA) task, where the subject's place was the

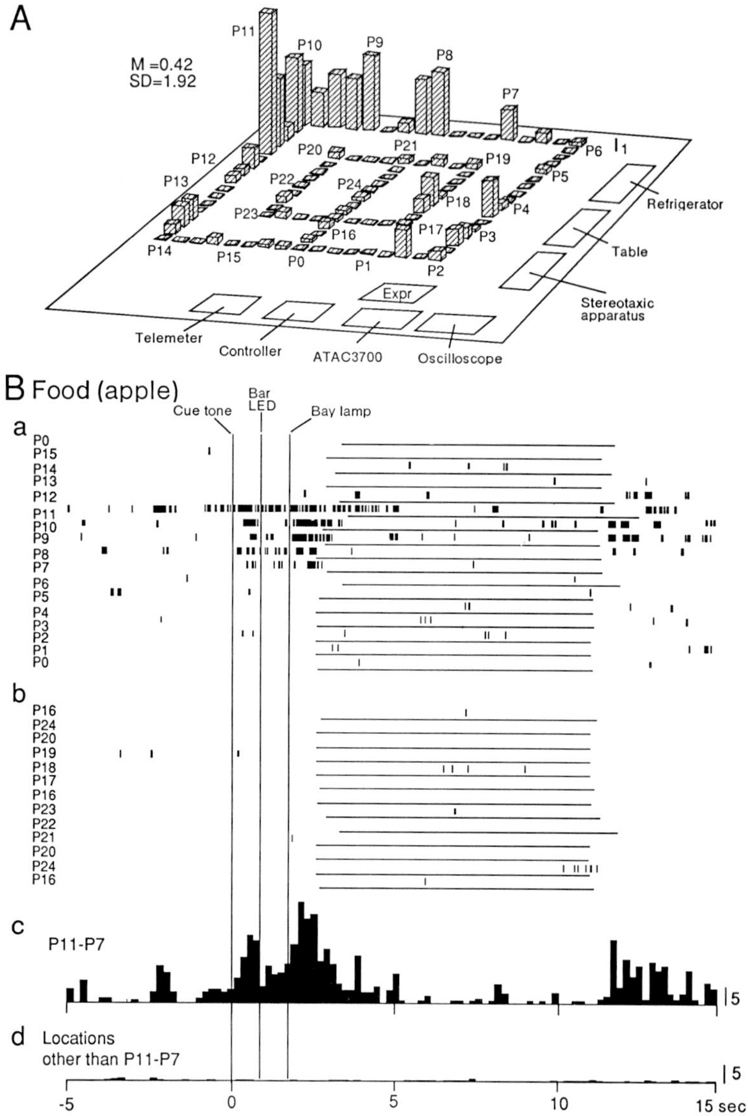

context to indicate behavioral attributes (go or nogo responses) of the objects, and the performance of the simple stimulus-response association (ISA) task, where each object indicated go or nogo response regardless of the subject's place (*3*). In the PCA task, each trial began when the monkey pressed and held a response bar. After holding the bar 4.0 to 6.0 sec (control phase), a pure tone was sounded to alert the monkey. After a 1.0 sec delay following the tone, the inside cab lamps were turned off for 3.0 sec (place phase). During the place phase, the monkey could see outside of the cab through the half-mirrored cab walls. The inside cab lamps were again turned on, and after another 1.0 sec delay an object (a red or blue block) behind the half-mirrored shutter was illuminated for 1.0 sec (object phase). When the object indicated go response based on the subject's location (Go trial), the monkey was required to release the response bar within 2.0 sec. If the monkey responded correctly, a juice reward was given after another 1 sec delay. The time course of a trial with a nogo response (Nogo trial) was similar, but the monkey was required to keep holding the bar for 5.0 sec. The behavioral attributes of the objects were conditional upon the locations of the cab. When the cab was located at P6 in Fig. 1C (Place I), a red block required the monkey to release the bar (Go trial), and a blue block required holding the bar (Nogo trial) to obtain the juice reward (symmetrical reinforcement). When the cab was located at P2 in Fig. 1C (Place II), the behavioral responses indicated by the two objects were reversed: the red block required holding the bar (Nogo trial) and the

←Fig. 3. Place dependence of task related neuronal response during a spatial moving task. A: neuronal responses during spatial moving task. Sequence descriptions as in Fig. 2. The neuronal activity was significantly high in locations P11, P10, P9, and P8. B: neuronal responses to spatial moving task. The raster displays show neuronal responses when the monkey was in the outer (a) and inner path (b) locations. Neuronal activity increases to cues, cue tones, apple, shutter opening and ingestion occurred when the monkey was located at P11, P10, P9, or P8. The neuron responded to cue tones and the sight of apple at P7; it did not respond to shutter opening or ingestion at P6. The two peristimulus histograms show a cumulative histogram (200 msec bins) of rasters in 6 trials from P11 to P7 showing high pretrial neuronal activity and further activity increase during the task (c), and a cumulative histogram of rasters in 18 other trials except in locations P11-P7 showing no neuronal response (d). Vertical bars in A, Bc, and Bd: spikes/sec.

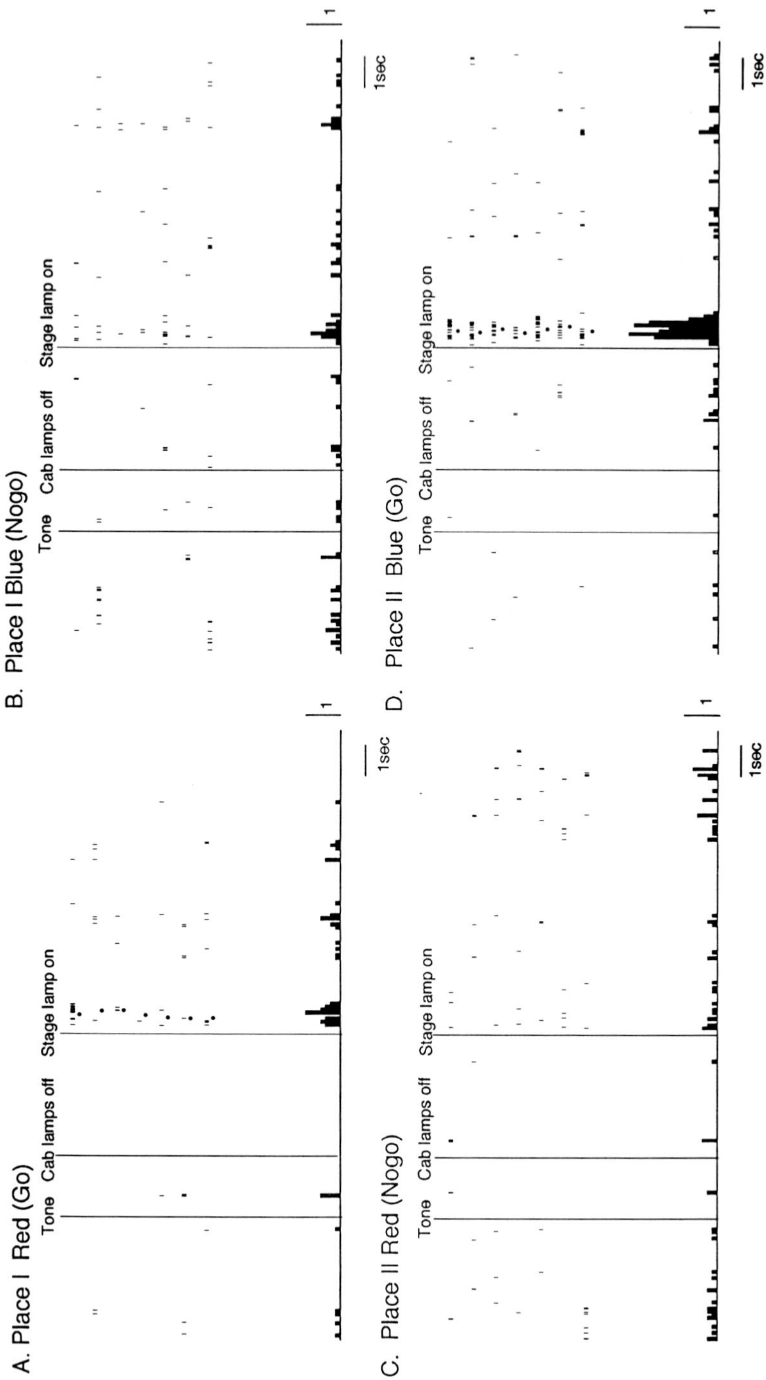
A. Place I Red (Go)
Tone
Cab lamps off
Stage lamp on
1sec
1
B. Place I Blue (Nogo)
Tone
Cab lamps off
Stage lamp on
1sec
1
C. Place II Red (Nogo)
Tone
Cab lamps off
Stage lamp on
1sec
1
D. Place II Blue (Go)
Tone
Cab lamps off
Stage lamp on
1sec
1

blue block required releasing the bar (Go trial) to obtain a reward.

The results demonstrated that some HF neurons responded preferentially in the PCA task, but not in the ISA task. Figure 4 shows an example of such HF neuron. This neuron specifically responded, during the object phase, to the blue block indicating Go response that was presented at Place II (D), but did not respond to other combinations of object-behavioral response-place (A–C). In the ISA task, this neuron did not respond to the same blue block that indicated Go response regardless of place (not shown). This indicates that the selective neuronal response to the combination of blue object and Go response in the PCA task was further constrained by the subject's location (Place II).

II. SIGNIFICANCE OF PLACE- AND CONTEXT-CORRELATES FOR MEMORY: A PERSPECTIVE

A series of studies in rats since the first report by O'Keefe and Dostrovsky (*6*) have reproduced place-correlates of the HF neurons during spatial navigation behavior such as maze learning. Also, as we have just shown, some neurons in the monkey HF had clear place-correlates. Moreover, such HF neuronal correlates were strongly influenced by the context in which the animal was. It should be noted that, in principle, such place-correlates of the HF neurons during ongoing behavior of animals are not necessarily attributed to memory. How do they correlate to the HF function in memory?

To consider organization of animal memory, here we would refer to an important report by Sakai and Miyashita (*10*). Using the stimulus-stimulus association task, they clearly showed some neurons in the ventral part of the anterior inferotemporal cortex (area TEav)

←Fig. 4. Responses of a typical HF neuron that responded to a specific combination of place, object, and behavioral response. A–D: rastergrams (upper) and cumulative 100 msec bin, 20 sec, 10 trial histogram (lower) of neuronal activity during trials of each place-object-behavioral response combination. Dots under each raster of Go trials correspond to bar release. Tone, onset of a pure tone; Cab lamp off, onset of spatial cue (outside view from the cab); stage lamp off, onset of presentation of an object (a red or blue block). This neuron responded in the object phase differentially to one specific combination; place II-Red block-Nogo response.

fired selectively to particular combinations of the stimuli (Fourier descriptors) which were explicitly associated in pairs in the task. These results should provide neuronal representation of long-term memory in the neocortex for stimulus-stimulus association. However, it should be emphasized that associations required in their task did not refer to the place or the context in which the animal was.

This point should be compared with the HF neuronal correlates to the place and the context. Consider various animal behaviors under various conditions. For example, consider condition A: a specific experiment in a particular room by a particular experimenter; and condition B: another specific experiment in another particular room by another particular experimenter. Here, we can recognize that there are two types of knowledge: one is that which remains constant under two conditions, and the other is dependent on the conditions, in other words, knowledge effective only under a specific condition. This dichotomy of knowledge has been proposed by many psychologists, and seems to be in accord with a classic theory in human memory by Tulving (*13*): so-called "semantic" and "episodic" memory. He proposed the encoding-specificity principle for episodic memory, in which not only specific information itself, but also the context in which the information is presented is simultaneously encoded into the engram.

In this inference, the place and/or the context might preferentially be processed in the HF, which is suggested to be one of the important brain structures for episodic memory. Consequently, the place and/or the context triggers a kind of selective response of the HF neurons. The gap between "memory" and "place-cells" might be filled by our framework. The early stage of learning about complex associations, which is essential for animals, presumably discriminates the context in which the animal is. This might explain place- and context-correlates of the HF neuronal activity since the place and/or the context is necessary to organize memory (see also refs. *3–5*).

SUMMARY

We showed the neuronal correlates in the monkey HF to place and context, which are comparable to those of the rat HF place cells.

Activity in the monkey HF neurons was analyzed during the spatial moving task guided by auditory and visual cues. Of 238 HF neurons analyzed, there were clear place-correlates in 79 neurons. Moreover, activities of some HF neurons were also strongly influenced by the context in which the monkey is. There have been no direct studies that link such place-correlates of HF neuronal activity recorded from ongoing behaving animals and the possible role of the HF in a class of memory suggested by several lesion studies. The results of our recent neurophysiological studies suggest that the HF is critical for organization of episodic memory that is essentially based on the animal's place and context.

Acknowledgments

This work was supported partly by the Japanese Ministry of Education, Science and Culture Grants-in-Aid for Scientific Research (05267103, 06454706, 06680786, 07244207, and 07558110), by the Uehara Memorial Foundation, and by the Human Frontier Science Program for the 3rd fiscal year.

REFERENCES

1 Cohen, N.J. and Eichenbaum, H. (1993). *Memory, Amnesia and the Hippocampal System*, Cambridge, MA: MIT Press.

2 Eichenbaum, H., Otto, T., and Cohen, N.J. (1994). *Behav. Brain Sci.* **17**, 449-518.

3 Eifuku, S., Nishijo, H., Kita, T., and Ono, T. (1995). *J. Neurosci.*, in press.

4 Kita, T., Nishijo, H., Eifuku, S., and Ono, T. (1995). *J. Neurosci.* **15**, 1683-1703.

5 Nishijo, H., Kita, T., Tamura, R., Eifuku, S., and Ono, T. (1995). This volume, pp. 17-30.

6 O'Keefe, J. and Dostrovsky, J. (1971). *Brain Res.* **34**, 171-175.

7 Ono, T., Nakamura, K., Fukuda, M., and Tamura, R. (1991). *Neurosci. Lett.* **121**, 194-198.

8 Ono, T., Eifuku, S., Nakamura, K., and Nishijo, H. (1993). *Neurosci. Lett.* **159**, 75-78.

9 Ono, T., Nakamura, K., Nishijo, H., and Eifuku, S. (1993). *J. Neurophysiol.* **70**, 1516-1529.

10 Sakai, K. and Miyashita, Y. (1991). *Nature* **354**, 152-155.

11 Scoville, W.B. and Milner, B. (1957). *J. Neurol. Neurosurg. Psychiatry* **20**, 11-21.

12 Squire, L.R. (1987). *Memory and Brain*, Oxford: Oxford University Press.

13 Tulving, E. (1983). *Elements of Episodic Memory*, Oxford: Oxford University Press.
14 Wilson, M.A. and McNaughton, B.L. (1993). *Science* **261**, 1055–1058.
15 Zola-Morgan, S.M. and Squire, L.R. (1990). *Science* **250**, 288–291.

BEHAVIOR

12

UNDERSTANDING THE BEHAVIOUR AND 'MINDS' OF OTHERS FROM THEIR FACIAL AND BODY SIGNALS: STUDIES OF VISUAL PROCESSING WITHIN THE TEMPORAL CORTEX

DAVID I. PERRETT, MIKE W. ORAM, ELISABETH WACHSMUTH, AND NATHAN J. EMERY

University of St. Andrews, School of Psychology, St. Andrews, Scotland, KY16 9JU, U.K.

This paper reviews studies of brain mechanisms which contribute to the perception and recognition of socially important signals that are manifest in the visual appearance of the face and the body. The aim is to provide an overview of cortical visual processing in the non-human primate, tracing the development of pattern sensitivity and illustrating how simple visual cues are used to establish brain cell responses that are sensitive to complex and abstract visual events.

Investigations into visual processing can be made by recording the activity of individual brain cells at different stages in the visual system of experimental animals. Since the work of Gross *et al.* (*15*), it has become clear that some cells in temporal cortex of the macaque monkey respond selectively to the sight of biologically important objects (such as faces and hands). The properties of such cells, therefore, offer a unique opportunity to study directly the brain mechanisms involved in processing complex visual patterns and their meaning.

I. THE STREAM OF PROCESSING SUPPORTING FACE RECOGNITION

To understand how cells in the temporal cortex become selective for specific patterns as complex as faces it is informative to examine the nature of visual processing occurring at earlier stages in the nervous system. On the basis of anatomical connections, the cerebral cortex responsible for the processing of visual information can be divided into many different areas. Recent studies of the macaque monkey distinguish at least 32 areas with visual functions (*13*). The interconnections of the areas (*50*) indicate two anatomical streams of processing, running dorsally and ventrally. The ventral stream appears to be responsible for the recognition of complex visual patterns such as faces and the realization of their meaning (*14, 21, 26, 46*).

Visual information flows from area V1 through a sequence of areas including V2, V4, posterior, central, and anterior inferior temporal cortex (PIT, CIT, and AIT, respectively) and the polysensory cortex within the anterior sections of the superior temporal sulcus (area STPa). Areas AIT and STPa are perhaps the final sites for the analysis of visual inputs before information is passed to limbic brain structures which are thought to be involved in memory and evaluation of emotional significance (*1, 2, 7*).

Physiologically, one can trace an increase in the complexity of visual function within the chain of visual areas in the ventral stream. Early studies of area V1 revealed cells selectively responding to elementary visual properties such as the orientation, contrast or colour of light at specific points in the retinal image (*18*). Specific populations of cells in the second and third visual areas (areas V2 and V3) respond selectively to the orientation of local contours defined by a variety of visual cues (*41, 47*). Having defined the boundary around a region of the image, an important subsequent stage in pattern recognition is to determine the shape enclosed by part or all of the image boundary (*26, 44, 45*). As the cells in the ventral stream are sampled at increasingly higher levels (V4, PIT, and AIT, *i.e.* an increasing number of synapses from the retinal input) there are two major changes in response properties. First, the cells respond to stimuli over an increasingly large region of the retinal image (*i.e.*, receptive field size increases) and second, the cells display selectivity for increasingly complex image shapes (*15, 45*). 'Elaborate' cells which are each selective for one stimulus shape, colour, or texture (*e.g.*, a green star shape or sharply tapered vertically striped region) are found in increasing numbers in V4 (2%), PIT (9%), and AIT (45%) (*45*).

II. CELL SENSITIVITY TO FACES AND BODIES

The elaborate cells within IT cortex provide an extensive alphabet of some 1,300 shape primitives which are used in various combinations to specify the presence of a great variety of more complex patterns such as faces and objects. The elaborate cells must also support the analysis of the subtle distinctions in facial appearance as different socially significant signals are emitted. The mechanisms through which intermediate levels of visual processing achieve these goals of establishing sensitivity to complex and subtly different patterns remain speculative (*26*), though some of the end points of visual analysis of facial information are apparent within area STPa.

In the cortex on the lower surface of the temporal lobes of the monkey brain (area AIT), Gross *et al.* (*15*) made a startling finding of cells with extremely complex selectivity for stimulus properties. For example, some cells appeared only to respond to the sight of a monkey's paw. Subsequent work in the STPa revealed populations of cells selective for the sight of faces (*8, 27*). While selectivity for shape is greatly increased in STPa cell responses, compared to those of cells earlier in the visual pathways, selectivity for other visual dimensions such as size, orientation, position, and colour is reduced. Thus the cells respond to many different instances of faces where the position of the face within the image, its size, orientation, and colour can vary (*8, 27, 28, 31, 43*).

Cell responses to faces have largely been interpreted as having a role in the recognition of faces as a class of objects or as individuals. There is evidence that sub-populations of cells responsive to faces are indeed sensitive to the differences between faces (*4, 12, 28, 33, 39, 51*). Considerations of the time-course of neural responses makes it clear that even though individual cells exhibit high degrees of selectivity in response discrimination, many cells with equivalent (high level) response selectivity are required to transmit information about any face or complex pattern between areas with maximal speed (*22*).

III. UNDERSTANDING THE DIRECTION OF ANOTHER INDIVIDUAL'S ATTENTION

The cells responsive to faces may not have an exclusive function in the recognition of individuals. Indeed, the selectivity of cells responsive to faces, particularly those within STPa, is consistent with these cells playing a role in the perception of social signals. One type of social signal that appears to be analysed extensively within the STPa is where another animal is directing its attention (*35*, *40*). We refer to these signals as social attention.

Different cells in the STPa are selective for different views of the head; some respond only to the front view, others only to the left profile view of the head, *etc.* (*16*, *39*). A cell maximally responsive to the face seen in left profile may thus signal that the attention of another individual is directed to the observer's left.

In many cases the direction in which another individual's head is pointing is not a reliable index of where that individual's attention lies. Gaze direction is a much better guide to the focus of another individual's attention and should therefore affect STPa cell responses. This prediction is borne out since most cells selective to a particular head view are also sensitive to gaze direction (*29*, *39*, *40*). Moreover, the direction of gaze and the head view to which cells are maximally sensitive tend to be aligned. For example, cells responsive to the head directed towards the observer (full face) are more responsive to eye contact than laterally averted gaze. By contrast, cells responding to the head turned away from the observer towards profile also respond more to laterally averted gaze. Similarly, in the vertical plane, many of the cells that are sensitive to the head pointing downwards are also sensitive to gaze directed down.

There are circumstances in which the direction of gaze of another individual is not clear, for example, when the eyes lie in shadow. In these cases, the direction of attention can still be analysed, though with less certainty, from the direction in which the head is pointing. Head angle thus provides useful backup information. Cells showing combined sensitivity to head view and gaze direction are therefore capable of signalling the direction of another individual's attention under a variety of viewing conditions (*29, 35, 40*).

Cells within the STPa have been found to process visual information about the body in addition to the face (*48*). Visual information arising from body cues appears to contribute to cell sensitivity in a way that is also consistent with the cells' role in analysing the direction of attention (*40*). For example, cells tuned to the left profile view of the head are often additionally tuned to the left profile view of the body (*48*). Such cell responses could contribute to signalling that an observed individual is attending in a direction to the observer's left. This signalling function could occur in situations where either the head or the body are partially occluded from sight.

Populations of cells sensitive to the vertical posture of the head often respond to the sight of a head pointing down and independently to the sight of gaze directed down. Some of the cells with this type of sensitivity were also found to be responsive to the sight of the body in quadrupedal but not bipedal posture. The visual sensitivity of this type of cell to the eyes, head, and body posture were each consistent with a role in coding the sight of an individual 'attending down' (*42*). For such cells different sources of visual information appear to have different priority in controlling responses; eye direction taking precedence over head posture and head posture taking precedence over body posture.

Cells conjointly sensitive to gaze direction, head view and body posture are capable of utilizing very different types of visual information, yet all of the information to which the cells are sensitive is compatible with the same conceptual interpretation of another individual's direction of attention. One can imagine that separate visual descriptions would be built initially for the appearance of the eyes, head and body and, at later stages in the analysis, the outputs of the appropriate versions of these lower level descriptions would be combined to establish selectivity for meaningful social signals (*42*).

IV. THE VALUE OF SOCIAL ATTENTION

In human society, gaze forms the basis for an extensive repertoire of communicative signals. Mutual gaze is a central component of bonding between individuals (*3*). Gaze direction can be an indication of mood state (*e.g.* downcast gaze of shame or depression). The direction of another's gaze also contains an explicit spatial reference within the environment. For example, within a monkey troop, upon hearing a predator alarm call it will be important to understand that X and Y (who may have given an avian predator alarm call) are both looking up and to the right. That direction specifies the location of potential predatory attack. In general, it is essential to comprehend the spatial reference to benefit from any visual or auditory communication.

The neuronal mechanisms coding head view and gaze direction could form the basis for perceiving where other individuals are attending. From these cues an observer can be made aware of objects of interest in the environment and the intentions of other individuals. Human infants (age 2 years) appear to comprehend the goal of the mother's gaze (*i.e.*, what the mother is looking at) and will direct their own gaze at the object fixated by the mother (*9*).

Baron-Cohen (*5*) has argued that the direction of gaze can provide clues to what someone else is thinking because it indicates what they are attending to. Understanding gaze direction could thus provide the seeds for a 'theory of mind' (ideas about what someone else is thinking). Such a capacity may be important for normal cognitive development in humans. Failure in understanding what other people are thinking or feeling has been suggested to be a cognitive deficit underlying the human developmental disorder of autism (*6, 20, 24*).

At the neuropsychological level there is evidence that damage to particular brain regions in monkeys and man can impair the ability to discriminate gaze direction while leaving other visual abilities intact (*11, 17, 32*). From neuropsychological case studies of human patients following brain damage it is clear that processing of different types of facial information can be dissociated (*49*). Some human patients have profound difficulties in identifying faces yet may recognize other attributes of the face such as expression, cues to speech, age and gaze direction without impairment (*10, 11, 49*). Judgements of gaze direction and social information about faces may thus involve different neural pathways from those involved in identification of individual faces (*11, 17*).

V. UNDERSTANDING FACIAL EXPRESSIONS

The visual analysis evident at the cellular level indicates that the temporal lobe is involved in far more than specifying the presence of a face in the retinal image. The neuronal analysis could support a rich understanding of the actions and possibly the 'intentions' of other individuals.

By analysing the visual information about the face in the context of information about the rest of the body and about the immediate environment, it is possible to establish a meaningful interpretation of the individuals an observer can see. For example, the specification that a face has an open mouth is useful as a starting point of visual analysis but it does not inform the observer whether the seen individual has an open mouth because they are vocalizing, eating or expressing a threatening gesture at the observer or another individual. Combining visual information from one facial feature with visual information from other parts of the body narrows down these alternatives and disambiguates the interpretations. To give a further example from the world of monkeys: a rounded open mouth (without teeth visible), lowered head angle, direct eye contact, raised eyebrows, and forward body lunges all combine to specify a threat directed at the observer. The same cues, but with the direction of gaze towards the observer's left, would also indicate a threat, but a threat directed at a different individual. Cells responsive to faces within the STPa again show sensitivity to multiple facial cues (including open mouth, raised eyebrows, head lowered, and eye contact) that can likewise constrain interpretation of the expressions of others (*25*).

In the monkey, area STPa provides a strong anatomical input to the amygdala (*2*). Through this pathway socially or emotionally significant aspects of facial and bodily information might be communicated to the limbic system. Indeed, the amygdala, like area STPa, contains substantial populations of cells sensitive to the sight of faces (*19*, *42*). It is therefore interesting to note that amygdala damage in humans may disrupt analysis of facial expression and discrimination of eye gaze direction but not the identification of faces (*1*).

VI. UNDERSTANDING BODY MOVEMENTS

The neuronal mechanisms described so far have involved the processing of static visual stimuli. Area STPa receives visual information about the motion of 'animate' objects from posterior motion processing areas. Comprehension of what other individuals are 'doing' may depend on the combined use of form and motion. Within the STPa there are many cell populations which appear specialized for the coding of specific limb or whole body movements (*23*, *30*, *36*, *37*). The cells appear to reflect combined sensitivity to both the form of the body and the nature of its movements. Collectively the cells can be thought of as providing a visual encyclopedia of recognizable actions of the face and body.

Again, the cell types can be arranged in a conceptual hierarchy starting with cells sensitive to the simple movements of single limbs and ending with cells sensitive to the complex patterns of articulation characteristic of whole body movements and cells sensitive to actions in which the body movements are related to objects or goals within the environment.

Cells sensitive to individual limb movements code articulations in specific directions towards or away from the observer or changing the limb elevation up or down. Some cells are responsive solely to particular head movements, others to leg or arm movements, and others to movements of smaller body or facial components (fingers, eyes, or mouth). All of these cell types display form sensitivity in that they respond specifically to one body part moving but not to equivalent movements involving a different body part (*e.g.*, arm movements but not leg movements). Form sensitivity is also reflected by the finding that the cells are unresponsive to control stimuli constructed to resemble the effective body part in size and shape (*30*, *36*, *37*).

However, the form sensitivity of some cells is sometimes reduced such that stick figures which articulate in the same way as the body or specific limb can evoke cell responses (*23*). Such sensitivity to patterns of articulation allow the cells to respond to a variety of body types with different colouration and patterning of fur or clothes (in the case of humans).

Many cells within area STPa have been found to show sensitivity for whole body movements made during walking (*30*, *36*, *37*). The majority of these cells are selective for both the direction of movement and the view of the body. For example, one cell might respond to the right profile view of the body moving to the observer's right but not to the left profile view moving in the same direction. Different cells are sensitive to body movements in different directions (left, right, towards, away, up, and down). These cells can represent a higher level of complexity as they respond to multiple cues arising from the walking motion.

VII. UNDERSTANDING ACTIONS

Direct evidence for cellular coding of actions has emerged from one population of cells within an area adjacent to STPa (on the lower bank of the STS). This population has been found to be sensitive to the sight of specific hand actions (*33*, *34*, *36*, *37*). Different subpopulations of cells were sensitive to different manual actions (picking, tearing, manipulating, holding). Actions differ conceptually from the facial gestures and signals discussed above because they are goal-directed. To interpret actions one needs to take into consideration both the agent of the action and the object or goal of the action. The goals of actions can usually be achieved via a variety of different routes. This means that the visual specification of an action must be more abstract since it needs to encompass the ways different versions of the actions achieve the same goal (*38*).

Cells responsive to hand actions code the movements of hands relative to the target or probable goal of the hand movements. For example, one type of cell described by Perrett *et al.* (*34*) responded to the sight of the experimenter's hand movements which brought an object towards the experimenter's mouth. Such cells remained unresponsive to the sight of a variety of hand actions (including reaching for, manipulating, picking, tearing an object). Hand movements carrying an object towards other parts of the experimenter's body produced less activity. Interestingly, movement of an empty hand towards the experimenter's mouth did not activate such cells. While the coding appears abstract, sensitivity to one visual relationship, a decreasing distance between the mouth and hand-held objects, may be all that is required to establish a crude visual representation of the action.

Thus, through a visual analysis of the face and body parts and relatively simple visual cues, cells in the temporal lobe can supply information which could enable the individual to realize that (a) a seen individual is bringing an item to its mouth, (b) that the individual perhaps 'intends' to bite the item, and (c) that the item in question is likely to be itself edible or to contain edible parts. Such a realization would be invaluable in guiding the actions of the observer in the hunt for food through social facilitation (individual X is drawn to forage where individual Y is seen to forage), or direct imitation (individual X copies the food processing seen exhibited by individual Y).

SUMMARY

Cells selectively responsive to the face have been found in several visual sub-areas of temporal cortex in the macaque brain. This paper reviews the way different cell populations may contribute to the recognition of face identity, face expression, and actions involving the face and body. As a general scheme of processing, the visual appearance of the face and body appears to be initially analysed in terms of independently mobile features (including the head, eyes, mouth, limbs, hands, and fingers). The orientation and motion of each of these body parts is first specified with respect to the viewer. These visual elements are combined in various ways to specify more complex whole body motions and conceptually linked social signals. Thus particular cell populations may signal 'social attention', *i.e.*, where other individuals are looking and directing their attention. This interpretation accounts for the combined sensitivity of some cells to gaze direction, head view and body posture. By relating the motion of the hands to the face, objects and locations in the world, a great variety of meaningful actions can be defined by cell responses. The visual specification of these actions could support an understanding of goal directed behaviour and inference of the intentions or minds of others.

Acknowledgments
This research was funded by project grants from the U.K. M.R.C., E.S.R.C., S.E.R.C., United States O.N.R. and Japan N.E.D.O. We thank L. Harrison, M.H. Harries, W. Dittrich, J.K. Hietanen, P.J. Benson, A.J. Chitty, A.J. Mistlin, and A.S. Head for their help in some of the experiments.

REFERENCES

1 Aggleton, J.P. (1993). *Trends Neurosci.* **16**, 328-333.
2 Aggleton, J.P., Burton, M.J., and Passingham, R.E. (1980). *Brain Res.* **190**, 347-368.
3 Argyle, M. and Cook, K. (1963). *Gaze and Mutual Gaze*, Cambridge, Cambridge University Press.
4 Baylis, G.C., Rolls, E.T., and Leonard, C.M. (1985). *Brain Res.* **342**, 91-102.
5 Baron-Cohen, S. (1991). In *Natural Theories of Mind*, ed. Whiten, A., pp. 233-251. Oxford: Blackwell.
6 Baron-Cohen, S. (1994). *Curr. Psychol. Cogn.* **13**, 513-552.
7 Brothers, L. and King, B. (1992). *J. Cogn. Neurosci.* **4**, 107-118.
8 Bruce, C.J., Desimone, R., and Gross, C.G. (1981). *J. Neurophysiol.* **46**, 369-384.
9 Butterworth, G.E. and Cochran, E.C. (1980). *Int. J. Behav. Dev.* **3**, 253-272.
10 Campbell, R. (1992). *Philos. Trans. R. Soc. Lond. B* **335**, 39-45.
11 Campbell, R., Heywood, C., Cowey, A., Regard, M., and Landis, T. (1990). *Neuropsychologia* **28**, 1123-1142.
12 Desimone, R., Albright, T.D., Gross, C.G., and Bruce, C. (1984). *J. Neurosci.* **8**, 2051-2062.
13 Felleman, D.J. and Van Essen, D.C. (1991). *Cerebral Cortex* **1**, 1-47.
14 Goodale, M.A. and Milner, A.D. (1992). *Trends Neurosci.* **15**, 20-25.
15 Gross, C.G., Rocha-Miranda, C.E., and Bender, D.B. (1972). *J. Neurophysiol.* **35**, 96-111.
16 Hasselmo, M.E., Rolls, E.T., Baylis, G.C., and Nalwa, V. (1989). *Exp. Brain Res.* **75**, 417-429.
17 Heywood, C.A. and Cowey, A. (1992). *Philos. Trans. R. Soc. Lond. B* **335**, 31-38.

18 Hubel, D.H. and Wiesel, T.N. (1968). *J. Physiol. Lond.* **195**, 215–243.
19 Leonard, C.M., Rolls, E.T., Wilson, F.A.W., and Baylis, G.C. (1985). *Behav. Brain Res.* **15**, 159–176.
20 Leslie, A.M. and Frith, U. (1988). *Brit. J. Dev. Psych.* **6**, 315–324.
21 Mishkin, M., Ungerleider, L.G., and Macko, K.A. (1983). *Trends Neurosci.* **6**, 414–417.
22 Oram, M.W. and Perrett, D.I. (1992). *J. Neurophysiol.* **68**, 70–84.
23 Oram, M.W. and Perrett, D.I. (1994). *J. Cogn. Neurosci.* **6**, 99–116.
24 Perrett, D.I. and Emery, N.J. (1994). *Curr. Psychol. Cogn.* **13**, 683–694.
25 Perrett, D.I. and Mistlin, A.J. (1990). In *Comparative Perception*, ed. Stebbins, W.C. and Berkley, M.A., vol. II, pp. 187–215. New York: John Wiley.
26 Perrett, D.I. and Oram, M.W. (1993). *Image and Visual Computing* **11**, 317–333.
27 Perrett, D.I., Rolls, E.T., and Caan, W. (1982). *Exp. Brain Res.* **47**, 329–342.
28 Perrett, D.I., Smith, P.A.J., Potter, D.D., Mistlin, A.J., Head, A.S., Milner, A.D., and Jeeves, M.A. (1984). *Hum. Neurobiol.* **3**, 197–208.
29 Perrett, D.I., Smith, P.A.J., Potter, D.D., Mistlin, A.J., Head, A.S., Milner, A.D., and Jeeves, M.A. (1985). *Proc. R. Soc. Lond. B* **223**, 293–317.
30 Perrett, D.I., Smith, P.A.J., Mistlin, A.J., Chitty, A.J., Head, A.S., Potter, D.D., Broennimann, R., Milner, A.D., and Jeeves, M.A. (1985). *Behav. Brain Res.* **16**, 153–170.
31 Perrett, D.I., Mistlin, A.J., Chitty, A.J., Smith, P.A.J., Potter, D.D., Broennimann, R., and Harries, M.H. (1988). *Behav. Brain Res.* **29**, 245–258.
32 Perrett, D.I., Mistlin, A.J., Chitty, A.J., Harries, M.H., Newcombe, F., and de Haan, E. (1988). In *Physiological Aspects of Clinical Neuro-ophthalmology*, ed. Kennard, C. and Clifford Rose, F., pp. 137–154. London: Chapman and Hall.
33 Perrett, D.I., Harries, M.H., Bevan, R., Thomas, S., Benson, P.J., Mistlin, A.J., Chitty, A.J., Hietanen, J.K., and Ortega, J.E. (1989). *J. Exp. Biol.* **146**, 87–114.
34 Perrett, D.I., Harries, M.H., Mistlin, A.J., and Chitty, A.J. (1989). In *Higher Order Sensory Processing*, ed. Guthrie, D.M., pp. 155–173. Manchester: Manchester University Press.
35 Perrett, D.I., Harries, M.H., Mistlin, A.J., Hietanen, J.K., Benson, P.J., Bevan, R., Thomas, S., Ortega, J., Oram, M.W., and Brierly, K. (1990). *Int. J. Comp. Psychol.* **4**, 25–50.
36 Perrett, D.I., Mistlin, A.J., Harries, M.H., and Chitty, A.J. (1990). In *Vision and Action: The Control of Grasping*, ed. Goodale, M.A., pp. 163–180. New Jersey: Ablex Publishing.
37 Perrett, D.I., Harries, M.H., Benson, P.J., Chitty, A.J., and Mistlin, A.J. (1990). In *AI and the Eye*, ed. Blake, A. and Troscianko, T., pp. 181–201. Chichester: John Wiley.
38 Perrett, D.I., Harries, M.H., Chitty, A.J., and Mistlin, A.J. (1990). In *Images and Understanding*, ed. Barlow, H.B., Blakemore, C., and Weston-Smith, M., pp. 94–107. Cambridge: Cambridge University Press.
39 Perrett, D.I., Oram, M.W., Harries, M.H., Bevan, R., Hietanen, J.K., Benson,

P.J., and Thomas, S. (1991). *Exp. Brain Res.* **86**, 159–173.
40 Perrett, D.I., Hietanen, J.K., Oram, M.W., and Benson, P.J. (1992). *Philos. Trans. R. Soc. Lond. B* **335**, 23–30.
41 Peterhans, E. and von der Heydt, R. (1989). *J. Neurosci.* **9**, 1749–1763.
42 Rolls, E.T. (1984). *Hum. Neurobiol.* **3**, 209–222.
43 Rolls, E.T. and Baylis, G.C. (1986). *Exp. Brain Res.* **65**, 38–48.
44 Sary, G., Vogels, R., and Orban, G. (1993). *Science* **260**, 995–997.
45 Tanaka, K., Saito, H., Fukada, Y., and Moriya, M. (1991). *J. Neurophysiol.* **66**, 170–189.
46 Ungerleider, L.G. and Mishkin, M. (1982). In *Analysis of Visual Behaviour*, ed. Ingle, D.J., Goodale, M.A., and Mansfield, R.J.W., pp. 549–586. Cambridge: MIT Press.
47 von der Heydt, R. and Peterhans, E. (1989). *J. Neurosci.* **9**, 1731–1748.
48 Wachsmuth, E., Oram, M.W., and Perrett, D.I. (1994). *Cerebral Cortex* **4**, 502–522.
49 Young, A.W. (1992). *Philos. Trans. R. Soc. Lond. B* **335**, 47–54.
50 Young, M.P. (1992). *Nature* **358**, 152–155.
51 Young, M.P. and Yamane, S. (1992). *Science* **256**, 1327–1331.

13

PERCEPTION OF GEOMETRIC ILLUSIONS BY PIGEONS, MONKEYS, AND APES

KAZUO FUJITA

Primate Research Institute, Kyoto University, Inuyama, Aichi 484, Japan

Recent advance of comparative studies of perception has suggested both similarity and dissimilarity between humans and nonhuman animals. For example, perception of geometric patterns was shown to be similar among pigeons, chimpanzees, and humans (*4, 20, 22*). In visual search performances, chimpanzees showed search asymmetry just like humans (*21*), whereas pigeons did not (*1*). The importance of these kinds of comparative studies of perception lies in understanding both the evolution of human information processing and the underlying neural mechanism.

Geometric illusions have captured much attention among psychologists because they are supposed to represent characteristics of human perception, and thus undoubtedly provide useful matcrials to discuss the evolution of visual information processing. However, very little is known about illusory perception by nonhuman animals. Knowledge about illusory perception of animals should have interesting and important implications to the models of human illusory perception because they are expected to suggest what is fundamental and what is derivative in human visual perception.

The distinction of homology and analogy is a very important

issue from this viewpoint. Though there have been several reports suggesting perception of several different geometric illusions by nonhuman species (*2, 3, 6, 15, 19*), they are little more than a simple demonstration of the illusory perception. A careful examination of the effects of a variety of parameters is essential in order to suggest homology of the same perceptual phenomena in different species.

In a series of studies, I investigated perception of the Ponzo illusion by pigeons (*Columba livia*), rhesus monkeys (*Macaca mulatta*), and chimpanzees (*Pan troglodytes*) (*10–13*). Following the initial demonstration of this illusory perception, I examined the effects of several figural parameters.

I. DEMONSTRATION OF THE PERCEPTION OF THE PONZO ILLUSION BY THREE NONHUMAN SPECIES

The subjects were placed in experimental boxes designed to test each species. A CRT display with a touch-sensor installed was located on one wall of each box. Personal computers presented black graphic patterns on the white background.

The task was common across subject species. A diagram of the trial is shown in Fig. 1. Trials started with a stimulus consisting of a horizontal target bar and the context patterns at the center of the display. Several "touch on" (or "peck at") this sample stimulus resulted in two choice locations at the two bottom corners of the display. One of the two locations was assigned to the "long" key and the other was assigned to the "short" key. Six different lengths of target were used and the subjects were trained to report "long" for the longer three lengths and "short" for the shorter three by responding on the appropriate choice locations. The longest bar was twice as long as the shortest bar. A piece of food was given to a subject at a probability differing across individuals when they responded on the correct location. A short doorbell sound or a brief flash of the display was given as a conditioned reinforcer for correct responses resulting in no food. A short timeout, during which the display went black, followed an incorrect choice.

The initial training was conducted using isolated bars without

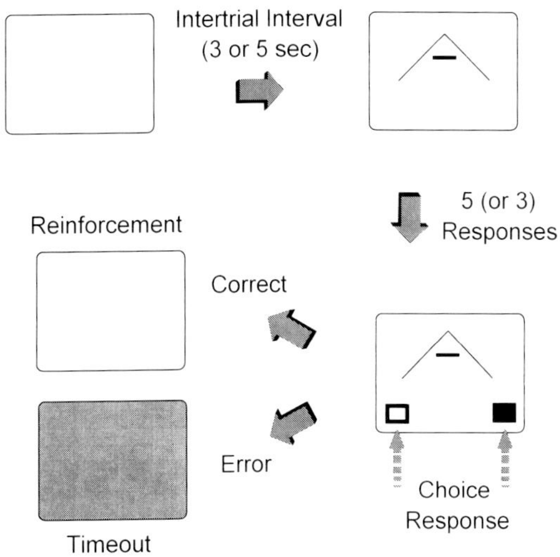

Fig. 1. A diagram of the trial.

any context stimuli. Context stimuli were superimposed after obtaining accurate performances on this initial training.

For the rhesus monkey and the chimpanzee, the context stimuli were an inverted-V, like the one appearing in Fig. 1. When consistent performances were obtained, the subjects were tested for their bias of reporting "long" and "short" when the relative locations of the target bar and the context patterns were changed. There were three conditions of test stimuli: high-context, middle-context (the training stimuli), and low-context. In the high-context stimuli, the upward-converging context appeared higher than the middle-context stimuli, thus making the apex of the inverted-V farther from the target bar. In the low-context stimuli, the inverted-V appeared lower than the middle-context stimuli, thus making the apex nearer to the target bar.

There was no timeout in the test trials; that is, all the choice responses resulted in either food or the conditioned enforcer. The test was done in a series of sessions interspersed by baseline sessions using training stimuli.

For pigeons, the context stimuli were initially 4 parallel vertical lines on each side of the target bar. When consistent performances were obtained, the subjects were tested for their bias of reporting in the same procedure as nonhuman primates. The context patterns of the test stimuli were 4 upward-converging lines on each side of the target bar. They were like 4 superimposed inverted-Vs having different top angles. As in the test for primates, there were three conditions of contexts, high-, middle-, and low-contexts. The testing procedure was the same as the one for primates.

Figure 2 shows the results of the test for the three species tested. The horizontal axis is the length of the target bar in graphic pixels.

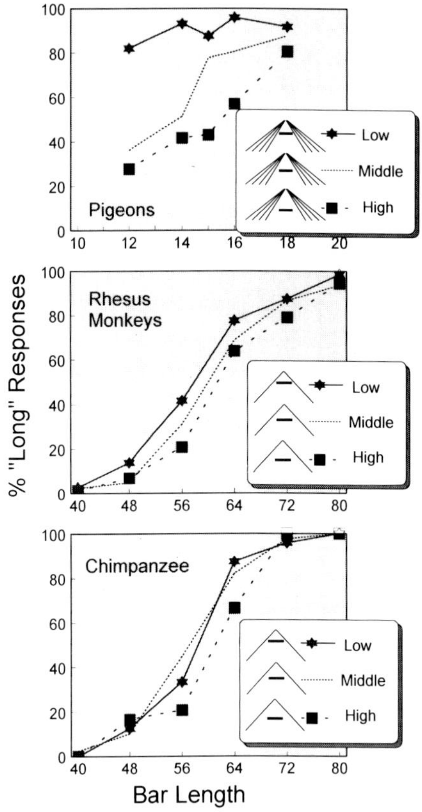

Fig. 2. Perception of the Ponzo illusion by pigeons (top: average of 3 birds), rhesus monkeys (middle: average of 3 monkeys), and chimpanzees (bottom: 1 chimpanzee). The horizontal axis is the length of the target bar in graphic pixels and the vertical axis is the percent of responses on the "long" key.

Note that the absolute lengths of the bars were different across species because of the difference in resolution and size of the monitor. The vertical axis is the percent proportion of the responses on the "long" key.

As is clear from the figure, the subjects tended to report "long" more often in the low-context stimuli, for which the apex of the inverted-V context approached the target bar, than in high-context stimuli. This tendency was common across the three species tested. The perception of the Ponzo illusion was thus demonstrated in all three species. The magnitude of the illusion, *i.e.*, the amount of the bias in reporting "long" or "short", was considerably larger in pigeons than in primates.

II. EFFECTS OF THE LINEAR PERSPECTIVE ON THE MAGNITUDE OF THE ILLUSION IN THE THREE NONHUMAN SPECIES

Effects of linear perspective on the magnitude of the illusion were also examined in the three species.

One classic account of this visual illusion is the "perspective theory" or "misapplied constancy scaling theory" proposed by Gregory (*14*). This theory proposes that humans unconsciously perceive perspective when they observe patterns like Ponzo figures. This induces constancy scaling compensation of the size of the patterns located on this perceived perspective; hence the overestimation of the size of the target bar near the apex of the inverted-V results. Actually, linear perspectives superimposed on the Ponzo figure were shown to enhance the illusion in some of the studies using humans (*17*, *18*). This theory appears to give qualitative support for some of the other illusions as well.

I prepared two types of context lines labeled perspective context and non-perspective context. The former had four upward-converging lines that converged into one point on each side of the target bar. This was the same type of context as the one used in the test trials for pigeons in the preceding section. The latter also had 4 lines on each side, but the lines were parallel to each other and had no converging point. The innermost lines were exactly the same between the two types.

The subjects were trained on the middle-context stimuli of both types of contexts. A series of test sessions followed after consistent performances were obtained. The testing procedure was the same as above.

Figure 3 shows the results. As in Fig. 2, all three subject species tended to report "long" more often for low-context stimuli than for high-context stimuli. This demonstrated once again perception of the Ponzo illusion by all the species tested. As before, the magnitude of the illusion was considerably larger in pigeons than in primates. However, for no species did there appear to be much difference in the magnitude of the illusion between perspective and nonperspective

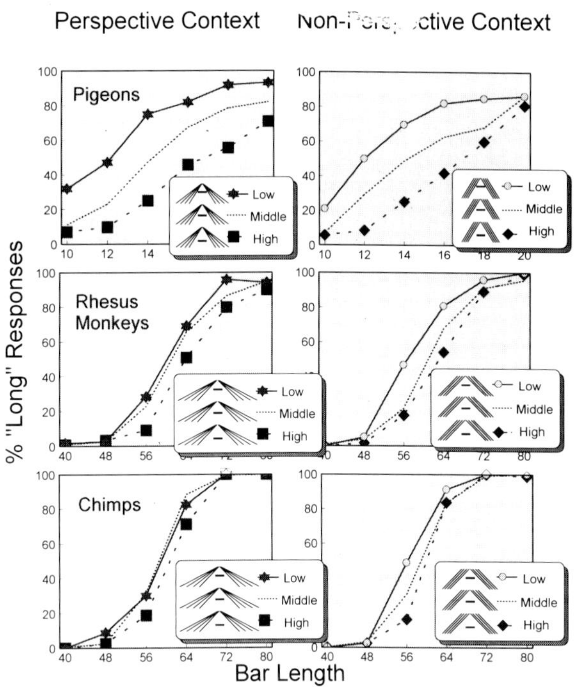

Fig. 3. Effects of the linear perspective on the magnitude of the Ponzo illusion in pigeons (top: average of 3 birds), rhesus monkeys (middle: average of 3 monkeys), and chimpanzees (bottom: average of 2 chimpanzees). The left 3 panels are for stimulus patterns with perspective context, and the right 3 panels are for those with non-perspective context. Others as in Fig. 2.

contexts. This result failed to support the "perspective theory" for the three nonhuman species tested here.

Interestingly, some of the human studies provide evidence against this theory too. For example, placing two vertical (instead of horizontal) bars in the inverted-V context leads to virtually no illusion (*16*). A texture analogue of the converging lines did not produce as good illusions as actual lines did (*7*). The present results were consistent with these studies.

III. EFFECTS OF THE GAP BETWEEN CONTEXT AND TARGET BAR ON MAGNITUDE OF THE ILLUSION IN RHESUS MONKEYS

Another potential source of the Ponzo illusion is the size of the gap between the target bar and the context lines (*8*, *9*). In humans, for example, two dots placed on either side of the target bar induces almost as good illusion as ordinary lines (*23*).

I prepared two types of contexts labeled "line context" and "dot context". The former was the same ordinary inverted-V as the one used in the previous section. The latter was two short vertical lines on either side of the target bar.

For line context stimuli, there were high-, middle-, and low-context conditions as before. For dot context stimuli, there were wide-, middle-, and narrow-context conditions. The size of the gap between

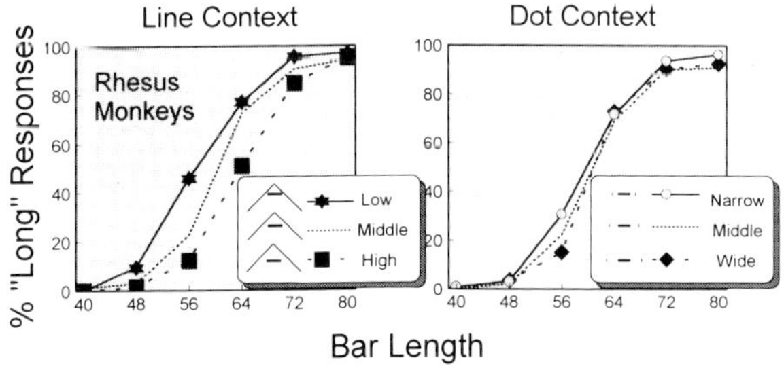

Fig. 4. Effects of the gap between the target bar and the context patterns on the magnitude of the Ponzo illusion in rhesus monkeys. The data are average of 3 monkeys. The left panel is for stimulus patterns with line context and the right for those with dot context. Others as in Fig. 2.

the context and the target bar corresponded in both types of contexts. The testing procedure was the same as before.

The results are shown in Fig. 4. Clearly, obtained biases were much larger for line contexts than for dot contexts, though there are biases for both types. Line contexts were more powerful in inducing the illusion. Thus the conclusions derived from this experiment may be that the Ponzo illusion by rhesus monkeys was only partly accounted for by the gap and that the contexts above or below the target bar also have some effect in enhancing this illusion.

IV. TEST OF HUMAN PERCEPTION OF THE PONZO ILLUSION

Humans were tested in a comparable procedure. There were five types of contexts: inverted-V (line context), perspective context, non-perspective context, dot context, and vertical-long-line context. The former four contexts were exactly the same as the one used to test nonhuman primates. The last one had two long vertical (therefore parallel to each other) lines instead of short lines in dot context. The vertical length of these long lines was the same as inverted-V. As was done for nonhumans, there were high-, middle-, and low-context

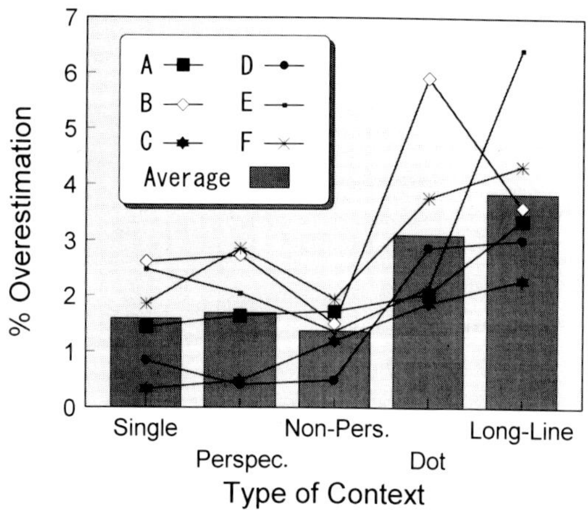

Fig. 5. The amount of the Ponzo illusion by humans for 5 different types of context patterns. The horizontal axis shows the type of context patterns and the vertical axis is the magnitude of the illusion in percentages. See text for other details.

conditions for the former three types of contexts, and there were wide-, middle-, and narrow-context conditions for the latter two.

The humans clicked the button on the computer mouse instead of touching the monitor. The trials were the same. After learning to report "long" and "short" for target bars presented alone, the subjects entered the test phase in which a titration was followed based on two successive "long" and "short" reports; *i.e.*, the length of the bar was increased when the subject reported "short" twice in consecutive trials of the same type of stimuli and vice versa. A total of fifteen (5 context types × 3 context locations) independent titration procedures were run and all the conditions of stimuli appeared randomly.

Figure 5 presents the averaged magnitude of the overestimation for each type of context. These values were calculated by subtracting underestimation for high- and wide-context conditions from overestimation for low- and narrow-context conditions compared with middle-context condition in a percentage. The data were based on the averaged bar length during the last 10 titrating trials for each stimulus. As is apparent, perspective context was not much more powerful than non-perspective context and inverted-V context. Dot context and vertical-long-line context were more powerful in inducing the illusion than the above three. The last result contrasted with those obtained in nonhuman animals.

V. CONCLUSIONS

The study series clearly demonstrated that nonhuman animals perceive the Ponzo illusion as humans do in a basic sense. The precise examination of the nature of this analogous perception has clarified several interesting similarities and differences across the species tested.

First, though the three nonhuman species similarly perceive the Ponzo illusion, the magnitude of the illusion was much larger in pigeons than in primates. This is probably due to a great difference in the visual system between the two taxonomic groups, but the precise cause will require future study.

Second, stronger linear perspective enhanced the illusion in no species including humans. For nonhuman species, this may be partly due to a difference in the ability to abstract perspective from line drawings. Difficulty in recognizing perspective from two-dimensional drawing was suggested at least in pigeons (*5*). However, this does not explain why humans obtained no gain in the magnitude of the illusion. Despite the previous data showing the positive effect of perspective (*17*, *18*), the contribution of linear perspective to the Ponzo illusion seems to be, at best, supplementary, minimal and situation-dependent.

Third, a very interesting difference was found between rhesus monkeys and humans in the effect of the short parallel lines (dot-context) to induce the illusion: dot-context induced stronger illusion than the converging lines in humans but the reverse was true in monkeys. This dissociation appears puzzling. However, both of the results commonly suggest that the gap between the target and the context does not account for the Ponzo illusion by itself. Contexts above or below the target may contribute to the Ponzo illusion, either positively or negatively dependent upon the identity of the species.

SUMMARY

Perception of the Ponzo illusion by pigeons, rhesus monkeys, and chimpanzees was studied. The subjects were trained to report the length of the horizontal target bar at the center of a CRT display as either "long" or "short" by touching one of the two choice locations corresponding to the length of the bar. A variety of context patterns were superimposed on the target bar. After accurate performances were obtained, the subjects were tested for their reporting bias when relative locations of the target and the contexts were changed. Humans were also tested for comparison using a titration of the bar length. The following major results were obtained. 1) All three species tended to report "long" more often as the apex of upward-converging lines approached the target bar. The perception of the Ponzo illusion was thus demonstrated. 2) Magnitude of the illusion changed in none of the species including humans when the strength of linear perspective of the context patterns changed. This failed to support the perspective account for the Ponzo illusion. 3) Whereas vertical short lines induced more illusion than upward-converging lines in humans, this was not the case for monkeys. These findings have implications for the mechanisms of geometric illusions and the evolution of visual information processing.

Acknowledgment

This study was supported by Grants-in-Aid for Scientific Research No. 04831005 and No. 06610069 from Japan's Ministry of Education, Science, and Culture.

REFERENCES

1 Allan, S.E. and Blough, D.S. (1989). *Percept. Psychophys.* **46**, 456–464.
2 Bayne, K.A.L. and Davis, R.T. (1983). *Bull. Psychon. Soc.* **21**, 476–478.
3 Benhar, E. and Samuel, D. (1982). *Anim. Learn. Behav.* **10**, 115–118.
4 Blough, D.S. (1985). *J. Exp. Psychol.: Anim. Behav. Processes* **11**, 261–280.
5 Cerella, J. (1977). *Pattern Recognition* **9**, 65–68.
6 Dominguez, K.E. (1954) *J. Genet. Psychol.* **85**, 105–127.
7 Fineman, M.B. and Carlson, J.A. (1973). *Percept. Psychophys.* **14**, 31–33.
8 Fisher, G.H. (1969). *Brit. J. Psychol.* **60**, 179–185.
9 Fisher, G.H. (1973). *Brit. J. Psychol.* **64**, 607–621.
10 Fujita, K. (1994). Paper presented at the 15th Bienn. Meet. Int. Primatol. Soc., August 6, Kuta, Bali, Indonesia.
11 Fujita, K., submitted.
12 Fujita, K., Blough, D.S., and Blough, P.M. (1991). *Anim. Learn. Behav.* **19**, 283–293.
13 Fujita, K., Blough, D.S., and Blough, P.M. (1993). *Anim. Learn. Behav.* **21**, 29–34.
14 Gregory, R.L. (1963). *Nature* **199**, 678–680.
15 Harris, A.V. (1968). *Redford Rev.* **22**, 61–72.
16 Humphrey, N.K. and Morgan, M.J. (1965). *Nature* **206**, 744–745.
17 Imai, S. (1982). *Tokyo Metrop. Univ. Jinmon Gakuhou* **152**, 1–18 (in Japanese)
18 Leibowitz, H., Brislin, R., Perlmutter, L., and Hennessy, R. (1969). *Science* **166**, 1174–1176.
19 Malott, R.W., Malott, M.K., and Pokrzywinski, J. (1967). *Psychon. Sci.* **9**, 55–56.
20 Matsuzawa, T. (1990). *Folia Primatol.* **55**, 24–32.
21 Tomonaga, M. (1993). *Percept. Motor Skills* **76**, 1287–1295.
22 Tomonaga, M. and Matsuzawa, T. (1992). *J. Comp. Psychol.* **106**, 43–52.
23 Yamagami, A. (1978). *Jpn. J. Psychol.* **49**, 273–279 (in Japanese).

14

INHIBITION AND FACILITATION OF VISUAL-MOTOR LINKS: IN RELATION TO ATTENTION AND MEMORY

SHINSUKE SHIMOJO*[1] AND YASUTO TANAKA*[2]

Department of Life and Cognitive Sciences, College of Arts and Sciences, The University of Tokyo, Tokyo, Japan[1] *and Weizmann Institute of Science, Rehobot, Israel*[2]

Spatial attention is an indispensable aspect of visual information processing: indispensable because the brain has a limited capacity whereas the constant flows of sensory inputs could theoretically be infinite. So, without selection and filtering by attention, it would be impossible to perceive what is important and to respond to it appropriately (*2*, *5*, *6*). On the other hand, some authors have recently argued that there is no limit to the capacity for visual information processing (*28*). Limited capacity may rather concern "selection for action" (*1*). For example, we can actively search for a particular object in a scene to reach for and grab. This view of spatial attention may be a part of the reason why reaction time (RT) has often been employed as a sensitive measure to access the selection and filtering processes, and it has revealed various effects in a spatio-temporal context.

In the literature regarding RT, however, there are two lines of evidence which are seemingly inconsistent in terms of direction of visual-attentional effect. One says that information processing at a previously cued location is facilitated, thus making a response faster there. The others says that visual attention is inhibited at a previous location so that a response is slower there. Our aim is to reconcile these two views and to integrate our understanding of facilitation and inhibition in visual-motor tasks.

I. INHIBITION OF RETURN

To begin with, consider a situation where a cue and a target are presented sequentially and randomly, either at the same or different locations. When the target is presented at the same location as the cue, RT is larger than that at a different location. This is so particularly when the interval between the onset of the cue and the target (stimulus onset asynchrony; SOA) is relatively large (300–1,500 msec; *15*). This has been called inhibition of return (IOR) and repeatedly duplicated not only in detection/button-pressing tasks (*3, 8, 11, 12, 23, 25*), but also in saccadic eye movements (*11, 16, 29*).

The underlying mechanism, however, is unknown. Since the reaction task involves various levels of processing, it is unclear at which level the effect occurs. There is a wide variety of stimulus parameters and tasks which have prevented us from more integrative understanding of "exogenous" (as opposed to "endogenous") spatial attention (*7, 8, 12, 14, 24*).

There have been at least two types of stimulus sequences with which the IOR has been observed. In the more classical type, a trial consisted of a cue and a target, whose locations were either the same or different (*14, 15*). The subject was asked to respond not to the cue, but to the target only. In the other type, there was no cue, and the subject simply had to respond to each of the targets which were presented sequentially and randomly (*8*). The RTs were then compared in terms of whether the location of one target was the same or different relative to that of the previous target. The IOR has been reported in this type as well as the classical cue-target types of experiment.

As a first step towards a more systematic investigation, we decided to choose the simpler design, that is, the no-cue type of stimulus sequence. We successfully duplicated the inhibition of return which had been reported by Kwak and Egeth (*8*) with a similar experimental design. Thus, the cue/target distinction is not a necessary condition for IOR, as originally suggested by Tassinari and his colleagues (*23*).

II. INHIBITION OF RETURN *VS.* FACILITATION OF RETURN: TASK DEPENDENCY

As mentioned, there is another line of evidence in the reaction-time literature suggesting that a previously cued location is facilitated, rather than inhibited in a subsequent response to it. Besides the wide variety of stimulus parameters and experimental designs, a notable difference was in the nature of the task. The subjects were asked for a simple detection or a saccadic eye movement in some of the studies, whereas in other studies they were asked to discriminate color, shape, or other features. In fact, some researchers suggested that the type of visual information, say, spatial *vs.* nonspatial attributes, was critical (*8, 23, 24*). To determine the critical conditions for inhibitory and facilitatory effects, we manipulated the nature of the task while maintaining the stimulus parameters, sequences, and the response as closely as possible.

Six subjects, including both naive and nonnaive, participated in the experiment. As in the pilot study, there was no cue. Location (top-left/top-right), color (red/green), and orientation of the target (vertically/horizontally elongated) were all independently randomized across trials, and presented on the CRT display controlled by a personal computer (Commodore Amiga 500). A typical stimulus sequence is illustrated in Fig. 1. Subjects were asked to fixate their eyes at the fixation point throughout the session, and to press a mouse button in response to each target presented. The interval between the button pressing response and the next target appearance (reaction stimulus interval: RSI) was randomized across trials (200/400/1,000/2,000 msec). Each of the subjects was engaged in four tasks in separate sessions: 1) simple detection, 2) location discrimination (left/right), 3) color discrimination (red/green), and 4) orientation discrimination (vertical/horizontal). The order of these sessions was randomized across subjects. Note that the stimuli and their sequence were identical across all the tasks. The subject conducted two alternative forced choices (2AFC) by button pressing for all the

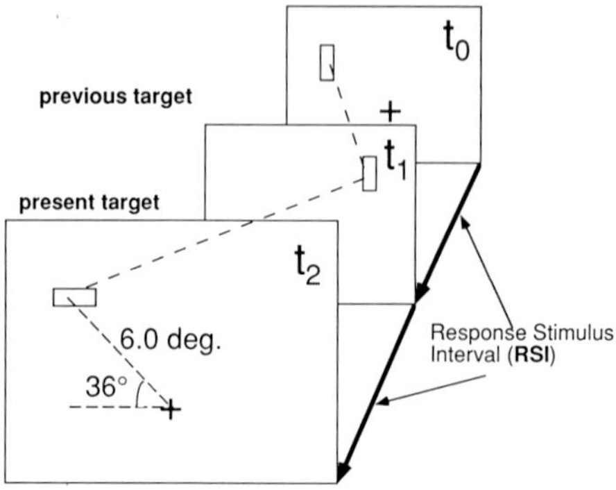

Fig. 1. Stimulus configuration and sequence for the RT experiment with a single target.

tasks, except for the simple detection task where the subject was asked to press a single button. Thus, the only difference was the nature of the task.

Target size was $0.5° \times 0.4°$, and luminance was 6.4 cd/m^2 (hue: red (.555,. 344), green (.320,. 555)). Size of the fixation point was $0.4° \times 0.4°$. Luminance of the background was 0.01 cd/m^2. The distance between the fixation point and the target was 6.0 deg. at the viewing distance of 114 cm. The subject's head was stabilized by a chin rest. Eye fixation was monitored in selected experiments using the *Ober2* system (Permobil Corp.). The experiment was conducted in a completely dark room, and the subject underwent a practice session (100 trials) before each session of the experiment. The total number of trials was 300 in each experimental session. RTs in error trials were eliminated from the data analysis. Sessions where the error rate for discrimination exceeded 10% were considered unreliable, and so were eliminated (see Tanaka and Shimojo (*22*) for more details of the method).

We analyzed the RT data based on the positional relationship of target between the previous and the present trials. In the results, a typical IOR was observed, that is, RT at the *same* location was *longer*, in the simple detection and the location discrimination tasks (see Fig. 2, top). The opposite pattern of results, however, was found in the color and the orientation discrimination tasks; RT at the *same*

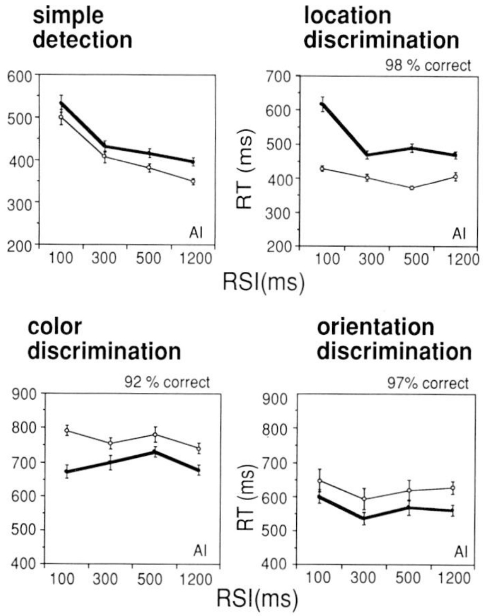

Fig. 2. RT as a function of RSI in the single target experiment. The results in four different tasks are shown in separate panels. Bold line indicates RT at the same location, and thin line indicates RT at a different location. Vertical bars indicate standard errors.

location as in the previous trials was *shorter* than that at a different location (see Fig. 2, bottom). We call this "facilitation of return (FOR)". The consistency of results across six subjects was reasonable, and statistical results (a four-way repeated measure ANOVA (RSI × location × color × orientation) supported them (*22*).

The opposite polarity of results between the two groups of tasks should be attributed to the task difference *per se*, because the stimulus parameters and the responses were identical. However, it was not the distinction between detection and discrimination, but rather the distinction in terms of information necessary to perform the task that seemed to be critical. On one hand, there were tasks which essentially require, or at least trigger, spatial orienting (detection and location discrimination). This is where we consistently found an Inhibition Of Return. On the other hand, there were tasks which require feature analysis (color and orientation discrimination), and this is where we consistently found Facilitation Of Return. The FOR is no doubt reminiscent of the benefits usually obtained with valid cues in the cost/benefit paradigm (*14*). The critical difference, however, was that here there was no cue/target distinction, nor any valid/invalid distinction.

III. GLOBAL ORIENTING *VS*. LOCAL FEATURE: ADDITIONAL FINDINGS

Follow-up studies in our laboratory have revealed a variety of facts about critical factors for inhibition and facilitation, all of which are informative to further characterize the underlying mechanisms. The findings will be summarized together with their implications in this section (see Tanaka and Shimojo (*22*) for more details).

1. Duplication in the Bottom Visual Field

All the data so far were obtained in the top visual field. Considering some anatomical speculation that the top and the bottom visual fields may be characterized by different neural projections, we repeated the same series of experiments in the bottom visual field. This time, two possible target locations were bottom-left and bottom-right of the fixation point. Within-subject comparisons of the results between two visual hemifields revealed highly comparable patterns: that is, the detection and the location discrimination tasks led to IOR while the color and the orientation discrimination tasks led to FOR. Although only two subjects have been completed to date the results indicate no systematic difference between the top and the bottom visual fields.

2. Vernier Discrimination Task

The dichotomy that we suggested is still ambiguous because spatial information could be used both for global orienting and to elaborate analysis of object features. The question is whether the *global* spatial orienting is critical, or any kind of *spatial* information is critical even at the finest scale, for IOR to occur. To answer this question, we next employed a vernier discrimination task. Two horizontally-aligned reference lines and a target dot between them were presented either at the top-left or the top-right of the fixation point. The subject was asked to judge whether the target dot was slightly misaligned with the reference lines upwards or downwards, and to press one of the two mouse buttons accordingly. The results indicate a clear pattern of FOR. This suggests that the distinction between two visual functions is better characterized as global orienting *vs.* local feature analysis, as opposed to spatial *vs.* nonspatial information.

3. Size and Luminance Discrimination Tasks

Size and luminance discriminations were also of interest for a similar reason. Since size is a spatial, yet local attribute, just as vernier, it could be categorized with the IOR-type or the FOR-type tasks (though the results of vernier discrimination are in favor of the latter). Since global orienting is often related to the so-called Magno pathway (*9*) which is characterized as being more sensitive to luminance-defined contrast than to color-defined contrast, it would be interesting to see if luminance discrimination is categorized with global orienting tasks or with local feature tasks. The target was presented again either at the top-left or the top-right, and the subject was asked to judge either the size (large/small) or the luminance (bright/dim) of it. The results indicated FOR in both tasks, again suggesting that the distinction is made between global orienting and local feature analyses, which include spatial/luminance features of objects.

4. Saccadic Eye Movement and Arm Reaching Tasks

There was still a question about the nature of the response. So far we had only used button pressing as a response. Could the IOR be generalized to other types of motor response? To investigate the robustness and generality of IOR, we employed more typical tasks of visually-guided orienting, saccadic eye movement, and arm reaching. The stimulus set for the saccade experiment was identical to that in the detection/button pressing task, except that the subject was asked to move his/her eyes as fast as possible to the target when it appeared. The *Ober2* system (Permobil Inc.) was used to record the eye movement. The temporal resolution of the measurement was set at 120 Hz, and the spatial resolution at a given condition was at or better than 0.3 deg.. The subject's head was stabilized by a chin rest and a biting board. The observation distance was set at 40 cm, thus the distance between the fixation point and the target was 17.1 deg.. The results showed a strong IOR. A four-way repeated measure

ANOVA (RSI × location × color × shape) revealed that the main effects of RSI and location were significant ($p < .01$).

We also tried an arm-reaching task, in which the subject had to control the cursor on the CRT display by moving a mouse on a flat surface so that the cursor would reach the target as quickly as possible. We then measured the performance time, which was defined as the time from the onset of target to the subject's button pressing at the target location. Here, we obtained the same but a somewhat exaggerated IOR (*20*).

Thus, IOR seems to reflect a common process among spatial orienting tasks (*14*). It is also consistent with a more specific hypothesis that the necessary and sufficient condition for IOR to occur is that a saccade be programmed, though not necessarily executed (*17*).

5. *Multiple Spatial Locations*

To further investigate spatio-temporal characteristics of the mechanisms underlying IOR and FOR, the target location was randomized across 7 possible positions while their distance from the fixation point was maintained constant. The distance of the present target from the previous target was measured retrospectively, and mean reaction times were calculated at each distance for the detection and the color discrimination tasks.

Results showed that mean RT at distance 0 was, in general, longer than at other locations in the detection, indicating IOR. The spatial range of IOR was rather large, however; it could be from 10 deg. to more than 20 deg. depending on the subject. ANOVA results indicated a significant interaction between RSI and distance, suggesting a highly transient characteristic of the IOR mechanism. On the other hand, FOR was obvious in the color discrimination because the mean RT at distance 0 was generally shorter than at other locations. The spatial range of FOR was much narrower, less than 6 deg. in all four subjects so far tested. ANOVA results indicated no interaction between RSI and distance, suggesting a sustained characteristic of the FOR mechanism.

6. *Duplication with Pop-out Displays*

One of the remaining questions was concerning the complexity of stimuli in relation to IOR. Let us now remember that the IOR is most likely to be related with sensory-guided spatial orientation. It has been commonly believed that the sensory-guided spatial orientation involves some sub-cortical loci such as the superior colliculus in its underlying neural circuit (*18*, *30*). Considering this, it is possible that IOR could occur only with the simplest type of stimulus, or "sub-cortically detectable" stimulus, such as a single target on a blank background. But on the other hand, the involvement of sub-cortical loci would not logically exclude possible involvement of cortical visual areas such as V1 and V2. This raises the natural question as to whether the IOR mechanism could extract the global location of a target when there are many visible distractors in the visual field (the stimulus situation well known as a "pop-out" display: *19*, *26*). This is a critical test because global orienting towards a popping-out target presumably requires local feature analyzers and their vigorous interaction at cortical levels.

Thus, we employed a pop-out display as shown in Fig. 3, which is similar to what had been employed by Maljkovic and Nakayama (*10*), and basically repeated the four tasks we had assigned with the single-target display (*21*). In this display, the target was defined by a color (red or green) which was different from that of distractors (green or red), and each diamond-shaped item was chopped off at either the top or the bottom. The location of the target (top-left or

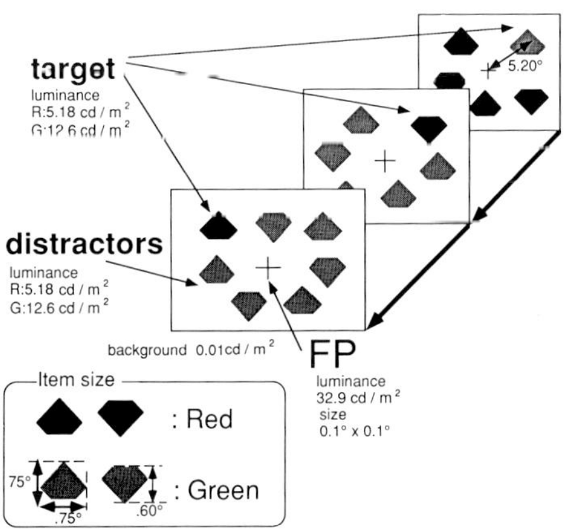

Fig. 3. The stimulus configuration and sequence for the pop-out experiments.

top-right), the locations of 15 distractors, the color (red or green), and the shape (top-chopped or bottom-chopped) of the target were all randomized across trials. The size of the target/distractor was 0.75×0.60 deg., the size of the fixation point was 0.1×0.1 deg., and the distance between the target and the fixation point was 5.2 deg.. The luminances of red and green were kept equivalent subjectively.

A summary of the results is shown in Fig. 4, where examples of typical results obtained from one subject were plotted separately for each task in each panel of the figure. The pattern of results was qualitatively the same as that with the single-target display: that is, IOR in the detection and the location discrimination tasks and FOR in the feature (color) discrimination tasks. Moreover, a closer comparison of Fig. 2 and Fig. 4 would reveal that the tendencies and dissociation are somewhat exaggerated with the pop-out display, even when the overall elevation of RTs is taken into account.

The results suggest that the IOR mechanism relies heavily on

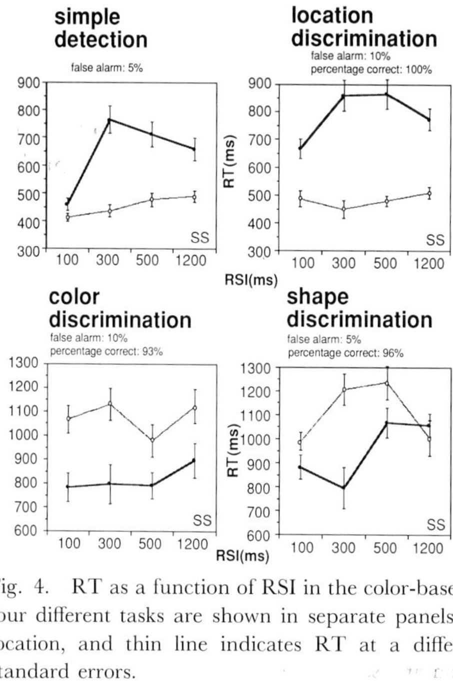

Fig. 4. RT as a function of RSI in the color-based pop-out experiments. The results in four different tasks are shown in separate panels. Bold line indicates RT at the same location, and thin line indicates RT at a different location. Vertical bars indicate standard errors.

cortical feature analyses, and that the dissociation between IOR and FOR should not be that between sub-cortical and cortical functions. Relationship between the dissociation we found in RTs here and the dissociations which have been advocated by Ungeleider and Mishkin (*27*) and Goodale and Milner (*4*) should be examined more closely in the future.

IV. CUMULATIVE REPETITION EFFECTS

We have so far only analyzed the data in terms of location. That is, we divided the trials into two categories retrospectively: those in which the target was presented at the same location as the previous one, and those in which the target was presented at the opposite location. Then, we compared mean reaction times of these two categories to find IOR and FOR. Thus, we have compared the trials with the same locations and the trials with opposite locations, regardless of the nature of the task. We could extend our analysis in two ways, by including repetitions of features as well as of location, and also by including repetitions over more than two trials (Tanaka and Shimojo, 1995).

All this could be done in the following way. For a particular visual attribute (location, color, or shape), we could first divide the trial sequence into "runs" with various run lengths defined by repetitions of the concerned attribute. For instance, if the target had been presented on the right side in three consecutive trials, and then switched to the left side in the next trials, this run had a length of three. Then, we could determine whether each trial was at the first place in the run, the 2nd, or the 3rd, and so on. After categorizing all the trials, we could calculate mean RT for each "order in sequence" category (1st, 2nd, 3rd, *etc.*). This could be done regardless of the nature of task, and the same analysis could be applied to feature repetitions as well. The number of trials included in the "order in sequence" category was largest in the 1st place, and became smaller in the 4th and 5th places, but anything beyond this was too small to be reliable. This was inevitable since all the stimulus parameters were randomized across trials.

Examples of the results are shown in Fig. 5. Here, all the subjects

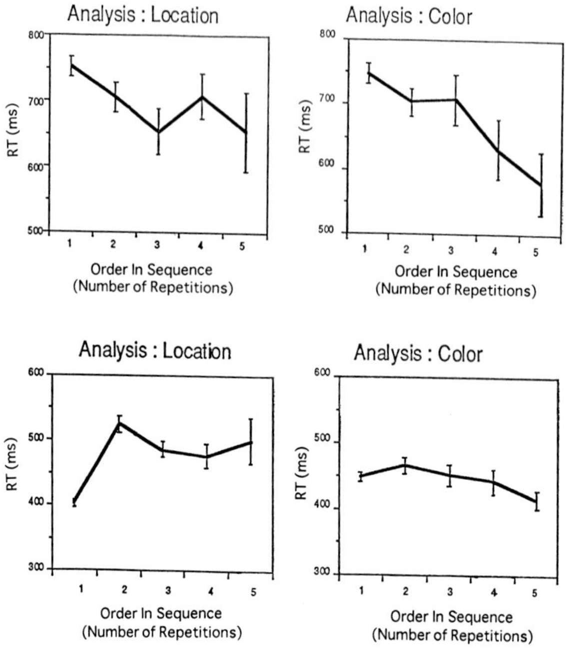

Fig. 5. Results of repetition analyses. The top two panels show the results in the color discrimination task, and the bottom two panels the results in the location discrimination task. The two left panels show the results of analysis by location repetition, and the right two the results of analysis by color repetition. In each panel, mean RTs are plotted against order in sequence, or number of repetitions. The vertical bars indicate standard errors.

tested were pooled for the analyses. The top two panels are for the color discrimination task, whereas the bottom two are for the location discrimination task. The left two are based on location repetitions, while the right two are based on color repetitions. Note that in the bottom left panel (location discrimination, repetitions of location), the mean RT in the 1st place in the sequence was shorter than that in the 2nd place. This is what we called IOR. Also note that in the top left panel (color discrimination, repetitions of location), the mean RT in the 1st place was longer than that in the 2nd place. This is what we called FOR. As is obvious by comparing these two panels, the facilitation turned out to be cumulative over 4 or 5 trials, whereas the inhibition was limited only to the next trial. We take this as another piece of evidence for the functional dichotomy between the inhibition and the facilitation.

Now look at the two panels on the right side of the figure. In one the task was color discrimination and the analysis was based on repetitions of color in the top right panel, while in the other the task was location discrimination and the analysis was based on repetitions of color in the bottom right panel. For the location discrimination, the color repetition had only a weak, if any, cumulative facilitatory effect (bottom right). For the color discrimination task, however, the color repetition had a remarkable facilitatory effect which was cumulative over at least 5 trials (top right). The saving in RT five-time repetition was more than 150 msec. There were three tendencies notable in the feature repetition results. First, the cumulative facilitatory effect had a longer lifetime over at least 4 or 5 trials. Second, this effect was mostly limited to the visual attribute which was relevant to the task. That is, repetition of a visual attribute had a cumulative effect primarily for the task on that particular attribute. Thus, the color repetition effect was clearest in the color discrimination task, whereas the orientation repetition effect was clearest in the orientation discrimination task. Third, color is somewhat dominant relative to orientation in terms of repetition effect. In fact, the orientation repetition effect was observed in the orientation discrimination task, but was not as notable as the color repetition effect in the color discrimination task. Also, there was a weak color repetition effect both in the location discrimination (bottom right panel) and the orientation discrimination tasks, but the orientation repetition effect was not found other than in the orientation discrimination task.

The feature repetition effects are reminiscent of the repetition effects which have been reported by Maljkovic and Nakayama (*10*) and Pashler and Baylis (*13*), although there were some notable differences. For example, Maljkovic and Nakayama employed a pop-out display (similar to ours) and a shape discrimination task, to obtain a cumulative color repetition effect. We are currently reanalyzing our data in the pop-out experiments with the four tasks for more direct comparison with their results.

SUMMARY

We found that IOR occurred when the task required global spatial information, regardless of the nature of the task, whether it was detection or discrimination. The IOR also occurred in different kinds of motor tasks such as saccadic eye movement or arm reaching.

We found the opposite tendency, *i.e.*, FOR when the task required identification of a detailed feature of the target, such as color, shape or veriner offset, even when the stimulus parameters were identical to those used before. This dissociation in reaction time may reflect a neurophysiological functional dichotomy between global orienting and local feature analyses.

We also found, somewhat surprisingly, the same kind of task-dependent dissociation when the pop-out display was employed. Even when the task-demanded information of global location was based on massively parallel feature analyses, it still led to inhibition, rather than facilitation. Thus, it seems likely that the neural pathway underlying the IOR significantly involves early cortical levels.

The IOR was not cumulative over 2 trials when the same location was repeated many times in the location discrimination task, whereas FOR could be accumulated over 5 or 6 trials at the same location in the repetition task. Cumulative facilitatory effects by feature (particularly color) repetitions were also found in the feature discrimination tasks. These results are consistent with the distinction between the spatial orienting and the feature analyzing functions, and multiple loci of visual short-term memory.

Acknowledgments

The research projects reported here have been partly supported by Grants-in-Aid for Creative Basic Research from the Ministry of Education, Science and Culture of Japan, the Human Frontier Science Program, and Nissan Cambridge Basic Research.

REFERENCES

1 Allport, D.A. (1989). In *Foundation of Cognitive Science*, ed. Posner, M., Cam-

bridge, MA: MIT Press.
2 Broadbent, D.E. (1958). *Perception and Communication*, Boston: Pergamon Press.
3 Gibson, B. and Egeth, H. (1994). *Percept. Psychophys.* **55**, 323–339.
4 Goodale, M.A. and Milner, A.D. (1992). *Trends Neurosci.* **15**, 20–25.
5 Helmholtz, H. (1910). *Hundbuch der Physiologischen Optik*, Hamburg: Verlag von Leopold Voss.
6 James, W. (1890). *The Principles of Psychology*, New York: Dover Publications.
7 Jonides, J. (1981). In *Attention and Performance*, ed. Long, J. and Baddelex, A., vol. IX, pp. 187–203. Hillsdale, NJ: Erlbaum.
8 Kwak, H.W. and Egeth, H. (1992). *Percept. Psychophys.* **51**, 455–464.
9 Livingstone, M. and Hubel, D. (1988). *Science* **240**, 740–749.
10 Maljkovic, V. and Nakayama, K. (1994). *Mem. Cogn.* **22**, 657–672.
11 Maylor, E. (1985). In *Attention and Performance*, ed. Posner, M.I. and Marin, O.S.M., vol. XI, pp. 189–204. Hillsdale, NJ: Erlbaum.
12 Nissen, J.M. (1985). In *Attention and Performance*, ed. Posner, M.I. and Marin, O.S.M., vol. XI, pp. 205–218. Hillsdale, NJ: Erlbaum.
13 Pashler, H. and Baylis, G. (1991). *J. Exp. Psychol.: Learn., Mem., Cogn.* **17**, 33–48.
14 Posner, M.I. (1980). *Q. J. Exp. Psychol.* **32**, 3–25.
15 Posner, M.I. and Cohen, Y. (1984). In *Attention and Performance*, ed. Bouma, H. and Bouwhuis, D.G., vol. X., pp. 531–556. Hillsdale, NJ: Erlbaum.
16 Posner, M.I., Rafal, R.D., Choate, L.S., and Vaughan, J. (1984). *Cogn. Neuropsychol.* **2**, 211–228.
17 Rafal, R.D., Calabresi, P.A., Brennan, C.W., and Scioltio, T.K. (1989). *J. Exp. Psychol.: Hum. Percept. Perform.* **15**, 673–685.
18 Robinson, D.A. (1981). In *Handbook of Physiology*, ed. Brooks, V.B., vol. II, Part 2, pp. 1275–1320, Bethesda, MD: American Physiological Society.
19 Sagi, D. and Julesz, B. (1985). *Science* **228**, 1217–1219.
20 Tanaka, Y. and Shimojo, S. (1993). *Invest. Ophthalmol. Vis. Sci. (Abstr.)* **34**, 1234.
21 Tanaka, Y. and Shimojo, S. (1994). *Invest. Ophthalmol. Vis. Sci. (Abstr.)* **35**, 1619.
22 Tanaka, Y. and Shimojo, S. (1995). *Invest. Ophthalmol. Vis. Sci. (Abstr.)* **36**, 902.
23 Tassinari, C., Aglioti, S., Chelazzi, Peru, A., and Berlucchi, G. (1994). *Vis. Res.* **34**, 179–189.
24 Terry, K.M., Valdes, L.A., and Neill, T. (1994). *Percept. Psychophys.* **55**, 279–286.
25 Tipper, S.P., Weaver, B., Jerreat, L.M., and Burak, A.L. (1994). *J. Exp. Psychol.: Hum. Percept. Perform.* **20**, 478–499.
26 Treisman, A. and Gelade, A. (1980). *Cogn. Psychol.* **12**, 97–136.
27 Ungeleider, L.G. and Mishkin, M. (1982). In *Analysis of Visual Behavior*, ed. Ingle, D.J.G. and Mansfield, J.W., pp. 549–586. Cambridge, MA: MIT Press.
28 Van der Heijden, A.H.C. (1991). *Selective Attention in Vision*, London: Routledge.
29 Vaughan, J. (1984). In *Theoretical and Applied Aspects of Eye Movement Research*, ed. Gale, A.G. and Johnson, F., pp. 143–150. Amsterdam: North-Holland.
30 Weiskranz, L., Warrington, E.E., Sanders, M.D., and Marshall, J. (1974). *Brain* **97**, 704–728.

15

COGNITIVE NEUROSCIENCE PERSPECTIVES ON AGING IN NONHUMAN PRIMATES

P.R. RAPP

Center for Behavioral Neuroscience, State University of New York at Stony Brook, Stony Brook, NY 11794-2575, U.S.A.

Research on the neurology of learning and memory has progressed at a remarkable pace over the last 15 years, culminating in a broad consensus concerning the regions of the brain that critically participate in mnemonic processing (*46*). Based on this background of information, recent efforts have engaged increasingly complex issues such as specifying the nature of memory representations supported by different regions of the brain, and defining how these neural systems interact in support of normal memory. Research on cognitive aging has advanced along parallel lines, capitalizing on experimental strategies and concepts that have fueled progress in the broader field of learning and memory research. The present chapter provides a selective review illustrating how a cognitive neuroscience perspective has informed and guided studies of aging in nonhuman primates. Evidence is outlined that memory-related processes supported by the prefrontal cortex are particularly susceptible to age-related decline. The status of learning and memory mediated by medial temporal lobe structures, in comparison, is highly variable across aged individuals. This more recent set of results has attracted considerable attention, yielding a potentially powerful strategy for identifying neurobiological alterations that account for variability in age-related cognitive decline. Against this background of substantial progress, converging evidence points to new avenues for establishing a more complete neuropsychological account of cognitive aging in the nonhuman primate.

I. TOWARD A NONHUMAN PRIMATE MODEL OF NORMAL COGNITIVE AGING

Normal aging in nonhuman primates is associated with a significant decline in cognitive processes supported by the prefrontal cortex. Early behavioral evidence consistent with this proposal derives from studies using the classic spatial delayed response (DR) task developed over 80 years ago (*19*). In the manual version of DR testing, a food reward is hidden in one of two recessed wells in a test apparatus while the subject watches. The rewards are covered with identical plaques and subjects are subsequently permitted to choose one of the locations, either immediately after baiting, or following a variable length retention interval. Animals are initially tested until they reach a criterion level of performance when little or no delay is imposed between the baiting and response phases of the trials, minimizing the memory demands of the task. Increasingly longer retention intervals are imposed during subsequent trials, typically ranging from 1 to 30 sec or more. Although procedural details differ widely across experiments, a common feature of spatial DR tasks is that accurate performance requires recall of a cued location from a relatively small number of options.

In an early series of investigations, Bartus and colleagues reported that rhesus monkeys in their late teens and twenties exhibit robust DR deficits (*e.g.*, ref. *9*). Specifically, aged monkeys performed as accurately as young subjects during testing with very short retention intervals, when demands on memory were minimal, but significant deficits emerged under conditions of increased memory load, at delays of 15 sec or more. Investigations using a wide variety of task variants have obtained broadly similar results, supporting the view that DR performance is particularly susceptible to aging in the monkey (*4, 5, 23, 26, 39*). Indeed, DR deficits are among the earliest behavioral signs of aging in the monkey (*5*), and are reliably observed in aged individuals that perform normally on other standard tests of learning and memory (*38*, *39*).

Considered against the substantial background of available evidence from lesion studies and electrophysiological investigations, these findings raise the possibility that DR deficits observed in aged monkeys partly reflect a decline in prefrontal cortical function. It has been known for many years that DR performance critically depends on the functional integrity of the prefrontal cortex (reviewed in ref. *18*). Unit recording studies in young monkeys support this view, demonstrating that the activity of a substantial population of neurons in the dorsolateral prefrontal cortex is selectively enhanced by cues presented in particular target locations, and that these cells often maintain increased firing rates following the offset of spatial cues, during the retention interval of a delayed saccade task (*15*). Thus, although there is clearly substantial functional heterogeneity across different subdivisions of the prefrontal cortex (*34*, *35*, *50*), compelling evidence has helped specify the role of these regions in normal DR performance.

A neuropsychological perspective leads to the prediction that aged monkeys might exhibit deficits on tasks other than DR that are known to be sensitive to prefrontal cortical dysfunction. A widely reported effect of frontal lobe damage, for example, is difficulty in altering behavior appropriately in response to changing task contingencies (*e.g.*, ref. *29*). In an early investigation, Bartus *et al.* (*8*) examined similar abilities in the aged monkey using a discrimination reversal task. Although the rate of discrimination learning tends to be more variable among aged individuals (*4*, *30*, *36*), as a group, older monkeys in this experiment learned new color and pattern discrimination problems as rapidly as younger animals. Significant impairments emerged, however, when the stimulus-reward contingencies were reversed. Similar to the perseverative responding noted in human patients with frontal lobe damage, reversal deficits observed in aged monkeys are at least partly attributable to a protracted period of continued responding to the previously reinforced stimulus. Although not all studies have replicated this effect (*36*), recent findings indicate that extensive behavioral testing can ameliorate perseverative tendencies in the aged monkey (*24*). Given that behavioral testing history has differed dramatically across experiments (*8*, *24*, *36*), this may be a significant factor contributing to the conflicting results on reversal learning in the aged monkey.

Additional behavioral evidence for a prefrontal cortical contribution to cognitive aging derives from studies that have manipulated demands on "executive system" functions that influence memory. For example, the effects of distraction on memory have been investigated using a variant of DR testing in which an irrelevant visual stimulus is presented during retention. Although this manipulation has no effect on performance in young subjects, accuracy in aged monkeys is substantially reduced under the same testing conditions (*7*). A recent report further demonstrates that age-related memory deficits are exacerbated as a function of demands on the temporal organization of memory (*39*). In this experiment, aged monkeys were significantly impaired when accuracy depended on memory for the order in which objects were presented. Recognition memory for the objects themselves, however, independent of presentation order, was normal in many of the same aged individuals (*38*, *39*). Source amnesia, or impaired memory for the temporal and spatial context in which information is acquired, is a prominent feature of human frontal lobe damage (reviewed in ref. *44*). Thus, from a neuropsychological perspective, this pattern of results is consistent with the proposal that the functional integrity of the prefrontal cortex is significantly compromised in the aged monkey.

II. INDIVIDUAL VARIABILITY IN BEHAVIORAL AND NEUROBIOLOGICAL AGING

For many otherwise healthy aged adults, the most troubling signs of aging occur, not in the domain of executive function supported by the prefrontal cortex, but instead involve declarative or explicit memory processes that critically depend on the functional integrity of medial temporal lobe and dienchephalic brain structures. The status of declarative memory, however, varies markedly across aged individuals. Variability in the effects of age on memory has been the subject of numerous recent reviews, yielding a convergence of evidence from studies in rats (*10, 16, 25*), monkeys (*37, 42, 48*), and humans (*31, 40, 45*). Rather than reiterate the same points here, the following sections provide a brief overview illustrating how selected investigations have capitalized on individual variability to explore potential neurobiological markers of memory dysfunction in the aged monkey.

A recent conference convened to survey the status of research on aging in nonhuman primates revealed that an increasingly common feature of experiment design is to combine behavioral and neurobiological assessment in the same subjects (conference transcript reported in *Neurobiology of Aging*, vol. 14). The value of this approach is that it provides a framework for identifying those neural alterations in the aged brain that are most closely associated with age-related impairments in learning and memory. By this strategy, the primary aim is to define neural alterations that are related to behavioral variability among aged individuals, as opposed to focusing on changes that are correlated with chronological age alone. Indeed, the perspective developed here is that measures reflecting the functional status of aged individuals can provide data necessary for appropriately interpreting the neurobiological consequences of normal aging. A recent investigation using quantitative autoradiographic methods, for example, reported that the density of adrenergic and serotonergic receptor subtypes in the prefrontal cortex was generally well-preserved in the aged monkey (*11*). Variability tended to be greater among aged individuals in dorsolateral prefrontal cortex (area 46), however, leaving open the important possibility that significant receptor changes might be present specifically among older monkeys that experience a robust decline in working memory processes dependent on this region of cortex. Consistent with this speculation, aged monkeys exhibit pronounced individual differences on tests of prefrontal cortical function in response to pharmacological manipulations of adrenergic systems (*3, 4*). Taken together, these considerations contribute to a growing recognition that chronological age, *per se*, may be a less informative and predictive variable in research on aging than analytic strategies aimed at defining the relationships between behavioral and neurobiological signature aging (see ref. *10* for a recent discussion of related issues).

Fundamental parameters such as neuron number provide a

logical starting point for examining the functional integrity of the aged brain. Studies in the monkey indicate that the effects of age on neuron number are regionally selective and, overall, subcortical nuclei appear more prominently affected than cortical structures. In a recent investigation using highly accurate and efficient stereological methods, no significant loss of neurons was observed in the hippocampal formation or layer II of the entorhinal cortex in aged monkeys (*49*). In addition, there was no apparent relationship, based on the small number of subjects available for analysis, between neuron number and the status of hippocampal dependent memory function in the aged individuals (Fig. 1). The same aged monkeys exhibited a substantial loss of medial septal nucleus cholinergic neurons that project to the hippocampus, however, averaging 40% fewer cells than young monkeys in rostral portions of the nucleus (*47*). Nonetheless, both the magnitude and regional selectivity of this loss were virtually identical in subgroups of aged monkeys distinguished by their performance on tasks sensitive to medial temporal lobe damage. Other studies have revealed a similar preservation of cortical neuron number in the frontal lobe of aged monkeys (*33*) against a background of substantial age-related loss in cholinergic and other neurochemically specific cells that provide widespread ascending projections (*21*, *43*). Identifying the mechanisms that confer selective vulnerability or protection on certain cell populations in the aged brain remains among the more significant challenges facing this field of research.

Considered together, these initial findings suggest that widely distributed neuron loss is unlikely to account for individual differences in memory function observed in the aged monkey. Other parameters reflecting the structural integrity of the aged brain, however, appear more tightly coupled to behavioral capacities. For example, although age-related declines in cholinergic cell number fail to predict the status of hippocampal memory function, increases in the size of surviving medial septal nucleus cells are observed selectively among aged subjects with impairments on memory tasks sensitive to medial temporal lobe damage (*47*). These results must be interpreted cautiously given the small number of subjects examined in this investigation. An intriguing possibility, however, is that age-

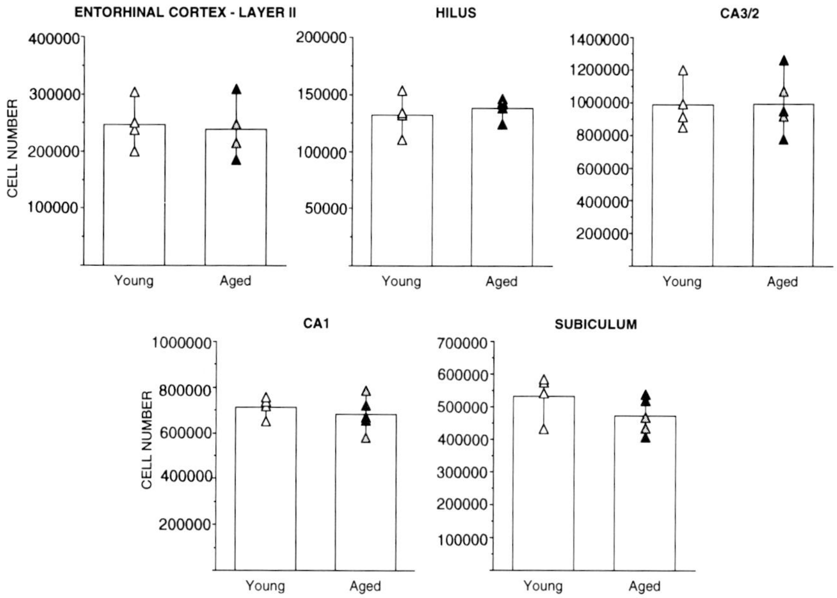

Fig. 1. Estimated neuron number in five subdivisions of the hippocampus for young ($n=4$) and aged ($n=5$) rhesus monkeys. Prior to histological evaluation, subjects were tested on a delayed nonmatching to sample test of recognition memory known to be sensitive to medial temporal lobe damage (see ref. *38* for details). Aged monkeys that scored below the entire range of values for the young group were classified as impaired, and the remaining aged subjects were classified as unimpaired. Neuron number was estimated using the optical fractionator method (*49*). Symbols represent the cell count values for individual subjects (young = open triangles, age-impaired = solid triangles, aged-unimpaired = shaded triangles). Neuron number was generally well-preserved in the aged monkey, and there was no indication of a disproportionate loss of cells in aged monkeys with documented memory deficits. Data adapted from West *et al.* (*49*).

related hypertrophy among cholinergic cells may be indicative of a reactive response. By this view, changes in cell size observed in memory-impaired aged monkeys might reflect the failure of protective or compensatory mechanisms, rather than signaling alterations that directly contribute to functional decline.

Other studies highlight changes of a degenerative nature associated with memory impairment in the aged monkey. In a recent investigation focusing on area 46 of the prefrontal cortex, chronological age and measures of cognitive status appeared unrelated to many morphological parameters including neuron density and the diameter of neuronal nuclei (*33*). In contrast, a subjective score of white matter abnormalities in the depths of the principal sulcus was highly correlated with a composite measure of memory function in a group of 5 aged monkeys. Future studies with larger sample sizes and objective quantitative methods will be necessary to establish the reliability of these results. The findings make a valuable contribution to a converging body of evidence, however, consistent with the suggestion that cortical neuronal loss is a less prominent feature of primate aging than previously thought, and that, among the morphological changes associated with advanced chronological age in the monkey, only a selective subset of alterations is related to individual differences in learning and memory function. This line of research in monkeys is in its infancy, and many important structural parameters (*e.g.*, dendritic architecture, terminal aborization, synaptic organization, *etc.*) have received only limited attention. Although substantial challenges remain, the studies briefly reviewed here emphasize that a valuable theoretical framework has emerged for establishing a detailed quantitative neuroanatomy of cognitive aging.

III. NEW HORIZONS FOR EXPLORING COGNITIVE DECLINE IN AGED MONKEYS

Learning and memory tasks that have been widely used to examine the effects of experimental brain lesions can provide a valuable window on the functional integrity of memory-related brain systems in aged subjects. Recent studies help define the limits of this approach suggesting that new experimental strategies will be necessary to develop more complete accounts of cognitive aging in the monkey. It has become increasingly clear, for example, that although standardized testing procedures like delayed nonmatching to sample (DNMS) are extremely sensitive to large medial temporal lobe lesions, more restricted damage confined to the hippocampus itself causes little or no impairment on many amnesia sensitive tasks (*1*, *32*). A reasonable interpretation of these findings is that the hippocampal formation (including the subicular complex and entorhinal cortex), together with closely associated cortical regions in the medial temporal lobe, comprises a highly integrated functional unit supporting declarative or explicit memory. By this view, damage limited to only one portion of the system (*i.e.*, the hippocampus) might be expected to cause relatively little memory impairment, particularly if other elements normally support similar processes and can therefore function in a compensatory fashion. This account, however, is difficult to reconcile with evidence in both rats and humans demonstrating that robust memory deficits can result from even subtotal damage restricted to the hippocampus (*20, 51*). Recent attention has focused instead on the possibility that DNMS and other testing procedures commonly used in nonhuman primate research might make only limited demands on the specific memory processing functions critically dependent on the functional integrity of the hippocampus (see refs. *12* and *13* for a related discussion).

A significant implication of these considerations is that standard learning and memory tasks might be expected to provide only coarse measures of the subtle changes that occur in hippocampal memory function during normal aging. Accordingly, a number of laboratories have begun to develop or adapt new behavioral testing procedures for exploring memory decline in the aged monkey. Anderson *et al.* (*2*), for example, evaluated spatial learning and memory using a procedure in which presentation of an array of 15 identical stimulus panels signaled the start of a trial. Each panel hid a food reward and maximum reinforcement was obtained by responding to a location exactly once in a test session, avoiding panels displaced earlier in the trial. Task acquisition was significantly impaired in aged monkeys, with the greatest number of errors occurring when the demands on memory for previous responses were greatest, late in each trial. Older monkeys also distributed their errors across fewer of the possible locations, perhaps reflecting a trend toward increased perseveration or behavioral rigidity with age.

Recent investigations have provided additional evidence that spatial information processing is compromised in the aged monkey (*22, 41*). Ongoing studies in our laboratory, for example, have examined spatial learning and memory in freely moving monkeys using a large open-field maze containing 8 reward locations distributed evenly around the perimeter of the apparatus. The ability to navigate through a testing environment may be an important feature of tasks that fully engage the spatial information processing functions of the hippocampus (*14*), and the present investigations were intended to incorporate this aspect of experimental design. In a version of testing adapted directly from parallel studies in rats, young monkeys rapidly learned to visit each reward location once per trial. Aged monkeys, in comparison, required substantially more training to acquire the task, and instances of successful performance often appeared to result from the use of alternate, non-spatial task strategies not seen in young monkeys, such as visiting reward locations in a fixed sequence. Additional data are needed to confirm that performance in normal young subjects is indeed supported by the spatial distribution of cues in the testing environment, and to establish that age-related deficits specifically reflect impaired learning and memory. Nonetheless, these preliminary observations are consistent with a much larger body of evidence from rats and humans demonstrating that spatial learning and memory are particularly susceptible to age-related deterioration.

Although tasks measuring learning and memory for spatial information are among the most sensitive procedures for documenting the effects of hippocampal damage and aging in the rat (*6, 17, 20*), many authors have suggested that spatial information processing is merely one example of a more general memory function supported by the hippocampus that is not limited to a particular category or class of stimulus information. A currently popular view, for example, is that the hippocampus selectively and critically participates in establishing representations in memory that reflect the relevant relationships between experienced items and events (*12, 13*). By this account, spatial learning and memory are exceptionally sensitive to hippocampal damage, not because the hippocampus plays a special role in processing spatial information, but instead because these tasks make heavy demands on the ability to learn and manipulate representations of the relationships between large numbers of cues in the testing environment. Thus, a prediction that follows from this proposal is that tasks emphasizing learning and memory for nonspatial relational information might provide sensitive measures of hippocampal dysfunction.

A recent experiment based on this perspective examined learning and memory for nonspatial hierarchical relationships in the aged monkey using a transitive inference procedure developed by McGonigle and Chalmers (*27, 28*). In this task, subjects are tested on a series of object discrimination problems such that the reward value of most of the stimuli depends on the object with which it is paired. For example, on one discrimination problem consisting of the simultaneous presentation of items A and B, only responses to A are rewarded. On other trials in which B appears with C, however, B is the correct and rewarded stimulus. In the present experiments, animals were trained on a series of 6 overlapping object discrimination problems: A+B−, B+C−, C+D−, D+E−, E+F−, F+G− (plus signs indicate the rewarded item in each pair). By this design, although items B through F are each rewarded on 50% of the trials, the relationship between the stimuli is accurately described by the hierarchy A>B>C>D>E>F>G. Memory for these hierarchical relationships was evaluated on probe trials in which novel, nonconsecutive pairs of items were presented for a response. On probe trials comprised of items B and F, for example, memory for the hierarchical relationships between items should yield preferential responding to B.

Despite the difficulty of this task, young and aged monkeys learned the instruction pair discriminations (*i.e.*, A+B−, B+C−, *etc.*) at a similar rate. Moreover, both groups exhibited substantial and equivalent levels of inferential responding on the various novel test pairs that can be formed from a hierarchical list of 7 objects (*e.g.*, CE, DF, BF, *etc.*). Thus, the findings essentially replicate earlier observations (*27, 28*) consistent with the suggestion that, under these testing conditions, monkeys learn and remember the objects according to their relevant hierarchical relationships. Although the absolute magnitude of this effect was similar across age groups, other aspects of performance suggest that this form of relational learning and memory is substantially impaired in the aged monkey. Response latency data proved particularly informative, revealing marked differences in young and aged subjects as a function of demands on relational information processing. Previous studies have recognized that the novel probe pairs can be distinguished according to their "symbolic distance", a measure reflecting the degree of disparity in

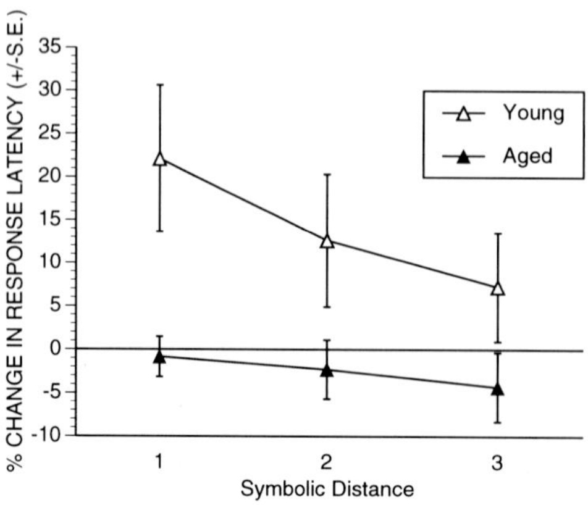

Fig. 2. Mean percent change in response latency as a function of symbolic distance for young ($n=12$) and aged ($n=5$) rhesus monkeys during probe testing on a transitive inference task (see text for details). Response latencies in young monkeys decreased substantially with increases in symbolic distance, but manipulating demands on relational memory had no effect in aged subjects. The percent change measure was calculated relative to each subject's latency on the instruction pair discriminations. Aged monkeys actually tended to respond more quickly on all trial types, indicating that the lack of a symbolic distance effect is not an artifact of slower baseline rates of responding in older animals.

hierarchical position of the items comprising the probe pairs. The CE probe, for example, has a symbolic distance of 1 because the stimuli are separated by 1 item in the series (*i.e.*, D). The BF pair represents the other extreme with a symbolic distance of 3 (*i.e.*, 3 missing items; C, D, and E). In the present experiments, response latencies in young monkeys decreased substantially with increases in symbolic distance (Fig. 2). Large disparities in hierarchical value are presumably recognized more rapidly than smaller differences, and the effect of symbolic distance on response latency in young monkeys is therefore consistent with the view that performance is supported by memory for these relationships (*28*). Response latencies in aged monkeys, by comparison, were largely insensitive to symbolic distance. Indeed, across a variety of different analyses, response latency signatures of relational memory processing were entirely absent in the aged monkey (*42*). Taken together with results from the spatial memory task, these findings suggest the procedures measuring learning and memory for relational information are particularly sensitive to aging in the monkey. Evaluating whether hippocampal dysfunction can account for these deficits remains an important topic for investigation.

SUMMARY

Studies on the neurology of learning and memory in young subjects have provided a valuable framework for developing a nonhuman primate model of normal cognitive aging. Behavioral research based on a neuropsychological perspective indicates that working memory and executive cognitive processes supported by the prefrontal cortex are substantially compromised during aging. Declarative memory function, by comparison, is less consistently affected across older individuals. These findings have prompted a productive line of investigation aimed at identifying age-related alterations in the hippocampus and anatomically related structures that might be responsible for variability in the status of declarative memory. Converging evidence suggests that cortical neuron loss is less pronounced in the aged primate brain than previously suspected, and that widely distributed cell death is unlikely to account for the memory deficits observed in many aged individuals. Against this background of considerable progress, a cognitive neuroscience perspective suggests that future advances will require new behavioral testing procedures for characterizing the nature and severity of age-related memory impairment. Recent theoretical considerations have proved valuable in this regard, guiding the development of tasks that emphasize precisely those memory processing functions that are thought to critically depend on the functional integrity of the hippocampus. Preliminary observations indicate that these new procedures are especially sensitive to aging, providing potentially powerful tools for establishing a detailed neuropsychological account of age-related memory decline in the nonhuman primate.

Acknowledgments

The author is indebted to David Amaral, Howard Eichenbaum, Mary Kansky, Jeffrey Roberts, Carmel Stanko, and Mark West for their contributions to the program of studies outlined in this chapter.

Original research and manuscript preparation was supported by NIH grants PO1 AG09973, RO1 AG10606, the CRPRC Base Grant (RR00169), and awards from NATO and the Human Frontier Sciences Program Organization.

REFERENCES

1 Alvarez, P., Zola-Morgan, Z., and Squire, L.R. (1995). *J. Neurosci.* **15**, 3796–3807.
2 Anderson, J.R., Anthouard, M., and de Monte, M. (1993). *Q. J. Exp. Psychol.* **46B**, 391–398.
3 Arnsten, A.F.T., Cai, J.X., and Goldman-Rakic, P.S. (1988). *J. Neurosci.* **8**, 4287–4298.
4 Arnsten, A.F.T. and Goldman-Rakic, P.S. (1985). *Ann. NY Acad. Sci.* **444**, 218–241.
5 Bachevalier, J., Landis, L.S., Walker, L.C., Brickson, M., Mishkin, M., Price, D.L., and Cork, L.C. (1991). *Neurobiol. Aging* **12**, 99–111.
6 Barnes, C.A., Treves, A., Rao, G., and Shen, J. (1994). *Sem. Neurosci.* **6**, 359–368.
7 Bartus, R.T. and Dean, D.L. (1979). *Exp. Aging Res* **5**, 385–400.
8 Bartus, R.T., Dean, R.L., and Fleming, D.L. (1979). *J. Gerontol.* **34**, 209–219.
9 Bartus, R.T., Fleming, D., and Johnson, H.R. (1978). *J. Gerontol.* **33**, 858–871.
10 Baxter, M.G. and Gallagher, M. (1995). *Neurobiol. Aging*, in press.
11 Bigham, M.H. and Lidow, M.S. (1995). *Neurobiol. Aging* **16**, 91–104.
12 Eichenbaum, H., Otto, T., and Cohen, N.J. (1994). *Behav. Brain Sci.* **17**, 449–472.
13 Eichenbaum, H., Otto, T., and Cohen, N.J. (1992). *Behav. Neural Biol.* **57**, 2–36.
14 Foster, T.C., Castro, C.A., and McNaughton, B.L. (1989). *Science* **244**, 1580–1582.
15 Funahashi, S., Bruce, C.J., and Goldman-Rakic, P.S. (1989). *J. Neurophysiol.* **61**, 331–349.
16 Gage, F.H., Chen, K.S., Buzsaki, G., and Armstrong, D. (1988). *Neurobiol. Aging* **9**, 645–655.
17 Gallagher, M., Gill, T.M., Baxter, M.G., and Bucci, D.J. (1994). *Sem. Neurosci.* **6**, 351–358.
18 Goldman-Rakic, P.S. (1987). In *Handbook of Physiology*, ed. Mountcastle, V.B., Plum, F., and Geiger, S.R., vol. 5, pp. 373–417. Baltimore: Williams & Wilkins.
19 Hunter, W.S. (1913). *Behav. Monogr.* **2**, 1–86.
20 Jarrard, L.E. (1993). *Behav. Neural Biol.* **60**, 9–26.
21 Kemper, T. (1993). *Neurobiol. Aging* **14**, 659–660.
22 Killiany, R.J., Lai, Z.C., Lai, J.C., Rosene, D.L., and Moss, M.B. (1994). *Soc. Neurosci. Abstr.* **20**, 388.
23 King, J.E. and Michels, R.R. (1989). *Anim. Learn. Behav.* **17**, 157–162.
24 Lai, Z.C., Moss, M.B., Killiany, R.J., and Rosene, D.L. (1994). *Soc. Neurosci.*

Abstr. **20**, 388.
25 Markowska, A.L., Stone, W.S., Ingram, D.K., Reynolds, J., Gold, P.E., Conti, L.H., Pontecorvo, M.J., Wenk, G.L., and Olton, D.S. (1989). *Neurobiol. Aging* **10**, 31–43.
26 Marriott, J.G. and Abelson, J.S. (1980). *Age* **3**, 7–9.
27 McGonigle, B.O. and Chalmers, M. (1977). *Nature* **267**, 694–696.
28 McGonigle, B.O. and Chalmers, M. (1992). *Q. J. Exp. Psychol.* **45**, 189–228.
29 Mishkin, M. and Manning, F.J. (1978). *Brain Res.* **143**, 313–323.
30 Moss, M.B., Rosene, D.L., and Peters, A. (1988). *Neurobiol. Aging* **9**, 495–502.
31 Nelson, E.A. and Dannefer, D. (1992). *Gerontologist* **32**, 17–23.
32 O'Boyle, V.J., Murray, E.A., and Mishkin, M. (1993). *Soc. Neurosci. Abstr.* **19**, 438.
33 Peters, A., Leahu, D., Moss, M.B., and McNally, K.J. (1994). *Cerebral Cortex* **4**, 621–635.
34 Petrides, M., Alivisatos, B., Evans, A.C., and Meyer, E. (1993). *Proc. Natl. Acad. Sci. U.S.A.* **90**, 873–877.
35 Petrides, M., Alivisatos, B., Meyer, E., and Evans, A.C. (1993). *Proc. Natl. Acad. Sci. U.S.A.* **90**, 878–882.
36 Rapp, P.R. (1990). *Behav. Neurosci.* **104**, 876–884.
37 Rapp, P.R. and Amaral, D.G. (1992). *TINS* **15**, 340–345.
38 Rapp, P.R. and Amaral, D.G. (1991). *Neurobiol. Aging* **12**, 481–486.
39 Rapp, P.R. and Amaral, D.G. (1989). *J. Neurosci.* **9**, 3568–3576.
40 Rapp, P.R. and Heindel, W.C. (1994). *Curr. Opin. Neurol.* **7**, 294–298.
41 Rapp, P.R., Kansky, M.T., Stanko, C.T., and Roberts, J.A. (1995). *Soc. Neurosci. Abstr.*, in press.
42 Rapp, P.R., Kansky, M.T., Roberts, J.A., and Eichenbaum, H. (1994). *Sem. Neurosci.* **6**, 369–378.
43 Rosene, D.L., Slattery, M., and Moss, M.B. (1992). *Annu. Congr. Int. Primatol. Soc.* **14**, 111.
44 Shimamura, A.P. (1995). In *The Cognitive Neurosciences*, ed. Gazzaniga, M.S., pp. 803–813, Cambridge: MIT Press.
45 Shimamura, A.P. (1994). *Sem. Neurosci.* **6**, 387–394.
46 Squire, L.R. (1992). *Psychol. Rev.* **99**, 195–231.
47 Stroessner-Johnson, H.M., Rapp, P.R., and Amaral, D.G. (1992). *J. Neurosci.* **12**, 1936–1944.
48 Walker, L.C., Kitt, C.A., Struble, R.G., Wagster, M.V., Price, D.L., and Cork, L.C. (1988). *Neurobiol. Aging* **9**, 657–666.
49 West, M.J., Amaral, D.G., and Rapp, P.R. (1993). *Soc. Neurosci. Abstr.* **19**, 599.
50 Wilson, F.A.W., Scalaidhe, S.P.O., and Goldman-Rakic, P.S. (1993). *Science* **260**, 1955–1958.
51 Zola-Morgan, S., Squire, L.R., and Amaral, D.G. (1986). *J. Neurosci.* **6**, 2950–2967.

16

SELF-MUTILATION BEHAVIOR AND BRAIN DOPAMINERGIC NEURON SYSTEM IN NEONATAL 6-HYDROXYDOPAMINE LESIONED RAT

HITOSHI OKAMURA*[1] AND CHIHIRO YOKOYAMA*[2]

Department of Anatomy, Kobe University School of Medicine, Kobe 650, Japan[1] *and Department of Anatomy, Kyoto Prefectural University of Medicine, Kyoto 602, Japan*[2]

Various types of clinical manifestations of self-injurious behaviors were encountered in a variety of diseases including character disorders, psychotic disorders, and mental retardation (*40*). Recent developments in neurochemical research suggest that impairment of neurotransmissions including opioids, dopamine, serotonin, and adenosine underlies these self-injurious behaviors (*40*), although none of the clinical trials to correct these impairments has been successful.

Self-injurious behavior is typically observed in patients with Lesch-Nyhan syndrome (*21*, *28*), a disease associated with an inborn deficiency of the purine-pyrimidine salvage enzyme hypoxanthine-guanine phosphoribosyl transferase (HPRT). Breese *et al.* (*5*, *7*) proposed neonatal dopamine-depleted rats by 6-hydroxydopamine (6-OHDA) as the model of Lesch-Nyhan syndrome, which manifests severe self-injurious behavior (hereafter called self-mutilation behavior (SMB)). In this article, we review our findings on the behavioral, neurochemical, and morphological characteristics of the neonatal 6-OHDA rat model of SMB, and further discuss the neuronal circuits underlying the behavior. Preliminary reports have been published in abstract form (*29, 44*).

I. DOPAMINE DEFICIENCY AND DISEASES WHICH MANIFEST SELF-MUTILATION BEHAVIOR

In classical Lesch-Nyhan syndrome, HPRT activity is less than 1%, and the retention of purine-pyrimidine metabolites occurs in most organs. The manifestation of symptoms of the central nervous system is thought to be related to the neuronal cell-death caused by the accumulated cytotoxic purine metabolites. Indeed, Page *et al.* (*30*) demonstrated that the reduction of HPRT enzymatic activity of fibroblast and the severity of central neurological symptoms were correlated in patients with Lesch-Nyhan syndrome. However, the association of the potential cytotoxic purine metabolites with the manifestation of the neurological symptoms is doubtful, since (1) no improvement or prevention has been reported, even normalization of the uric acid serum concentration by alloprinol treatment from the neonatal period (*26*); (2) some patients lack HPRT activity completely but do not manifest neurological symptoms (*14*); (3) HPRT-gene deleted mutant mouse did not manifest neurological symptoms of the central nervous system (*19*). These findings suggest the existence of a "primary factor" important for the unique neurological symptoms other than the abnormality of purine-pyrimidine metabolism.

Although various speculations have been made, it is now clear that the impairment of dopaminergic neurotransmission is closely related to this primary factor. Using the postmortem brain, Lloyd and his collaborators (*23*) found that dopamine levels are reduced by 70–90% in Lesch-Nyhan patients. Clinical studies on Toulet syndrome, in which SMB is one of the main symptoms, also found that the dopamine nervous system was impaired (*33*). A neonatal 6-OHDA dopamine lesioned animal model of SMB which lacks deficiency of HPRT activity (*5*) also supports the hypothesis. In primates, Goldstein *et al.* (*18*) succeeded in making SMB by the electric lesion of midbrain dopaminergic neurons. These findings suggest that the impairment of dopaminergic systems is related to SMB. Interestingly, a recent report by Jinnah *et al.* (*20*) demonstrating the reduction of dopamine level in HPRT-gene deleted mutants suggests that a single gene mutation of HPRT deficiency could impair the dopamine neurotransmission. Discovery of a large amount of adenosine A_2 receptor in the striatum (*17*) may also link dopamine impairment and purine metabolism impairment through adenosine receptors.

II. INTRACRANIAL 6-OHDA TREATMENT TO NEONATAL RATS

In 1984 Breese and his collaborators (*5*) found that neonatal 6-OHDA lesioned rats showed SMB following intraperitoneal injection of L-dihydroxyphenylalanine (L-DOPA) as adults. Comparing the behavioral characteristics of the neonatal 6-OHDA lesioned animal and those of Lesch-Nyhan disease, they proposed the neonatal 6-OHDA lesioned animal as a neurochemical model of the susceptibility for self-mutilation and the motor dysfunction observed in the Lesch-Nyhan syndrome (*4, 10*).

To make SMB rats, we basically followed the experimental protocol of Breese *et al.* (*6*). Progeny of pregnant Wistar rats were first given desipramine (20 mg/kg, i.p.) to protect the noradrenergic neurons, and then 6-OHDA (100 μg free base/5 μl in saline containing 0.1% ascorbic acid) intracisternally on postnatal day 1 and day 3 under deep cold anesthesia (*44*). In this protocol, we adopted two injection schedules to avoid the technical failure of 6-OHDA injection into the cisterna magna, and to deplete the dopamine content more completely. The smaller dose injected (10 μg) did not induce SMB, and a higher injected dose (200, 500 μg) increased the death rates due to the 6-OHDA non-specific toxicity.

However, the incidence of L-DOPA-induced SMB did not reach 100% even in accurately 6-OHDA injected neonates. Does this suggest the existence of a genetic difference in this background? We cannot answer this question at present, but since the degree of dopamine depletion is different even using the same schedules, and there is a correlation of dopaminergic neuron deficiency rate and the development of SMB, if a genetic background exists it would be correlated to the vulnerability of dopaminergic neurons to the toxicity of 6-OHDA. Furthermore, young rats injected at postnatal days 10 and 12 at a higher dose (150 μg of 6-OHDA; the dose was determined considering the increase in brain weight during 10 days), or adult rats at the same dose, did not show SMB following L-DOPA loading, although they showed the increase of locomotor activity following this challenge. This suggests that the induction of SMB is related to the timing of the dopamine-lesioning. Thus, whether SMB occurred or not was determined by the difference in maturity of the brain in each individual rat.

Other injection routes of 6-OHDA to destroy the midbrain dopaminergic neurons can also make SMB rats. We succeeded in producing SMB rats following bilateral lateral ventricular injections (25 μg each), ipsilateral lateral ventricular injection (50 μg or 100 μg), or injections to the third ventricle (50 μg).

III. BEHAVIORAL CHARACTERISTICS OF NEONATAL DOPAMINE-LESIONED RATS

Although over 90% of dopaminergic neurons were destroyed by the neonatal 6-OHDA treatment as processed above, they did not show somatomotor disturbances such as rigidity or akinesia as observed in adult dopamine lesions rats by 6-OHDA (*7*, *11*, *34*). Their motor behavior was normal in an untreated condition. However, L-DOPA injection induces a tremendous behavioral change in these rats. In our experiment, we evaluated their behavior at the age of 4 to 6 weeks. After 30 min of pretreatment of peripheral aromatic L-amino acid decarboxylase (AADC) inhibitor, Ro4-4602 (50 mg/kg, i.p.), animals were loaded with L-DOPA (100 mg/kg, i.p.). 6-OHDA-treated rats were divided into those with and without SMB occurrence, SMB(+) and SMB(−), respectively. The dose of L-DOPA was that according to Breese *et al.* (*5*), in which they found that L-DOPA loading from 100 mg/kg to 1,000 mg/kg did not change the incidence of SMB, but the reduction from 100 mg/kg to 10 mg/kg could reduce this incidence. This suggests that 100 mg/kg is sufficient to divide the rats into SMB(+) and SMB(−).

SMB was defined as the constant self-biting that caused a break in the skin. Once SMB started, the rats continued biting without looking aside, until SCH-23390, a D_1 antagonist, was injected into their peritoneum. The rats 'mutilate' their own skins so vigorously that many of them bite not only skin, but also muscles and bones, and sometimes all fingers and tails are lost if they are allowed to do so.

The beginning of SMB differs in each rat from 20 to 90 min. The maximal induction time is 30–60 min after the L-DOPA loading. The relatively predominant regions bitten were upper extremities, lower extremities, and abdomen (Fig. 1). Some rats bite their own tails, abdomen, and chest. Interestingly, the biting regions were not consistent in one rat. One rat bit his right upper extremity at the age of 4 weeks, and his abdomen at the age of 4 months.

The biting targets were their own body, not other animals. This is different from the attack behavior such as is caused by muricide. When exposing mice to these neonatal 6-OHDA rats exhibiting SMB, rats paid no attention and continued to bite their own skins eagerly.

SMB is not a temporary phenomenon. We found most rats which showed SMB at 4 weeks demonstrated SMB at 4 months and 12 months of age following L-DOPA loading. This suggests that the SMB is induced by the permanent alteration of neuronal transmission of the brain.

L-DOPA was converted to dopamine in the brain, and induced

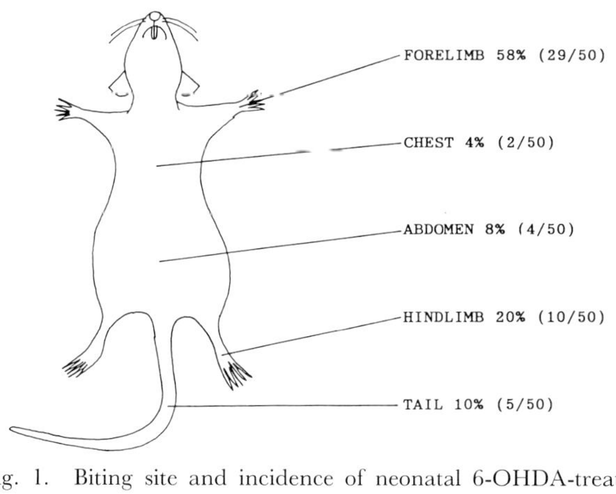

Fig. 1. Biting site and incidence of neonatal 6-OHDA-treated rats when loaded by L-DOPA.

not only SMB but also various behavioral changes through the dopamine receptor. To determine the correlation of hyperactivity and stereotypy, their behaviors were checked in a clear plastic cage (30 × 50 × 54 cm) on Animex for 60 min (Fig. 2). Following L-DOPA

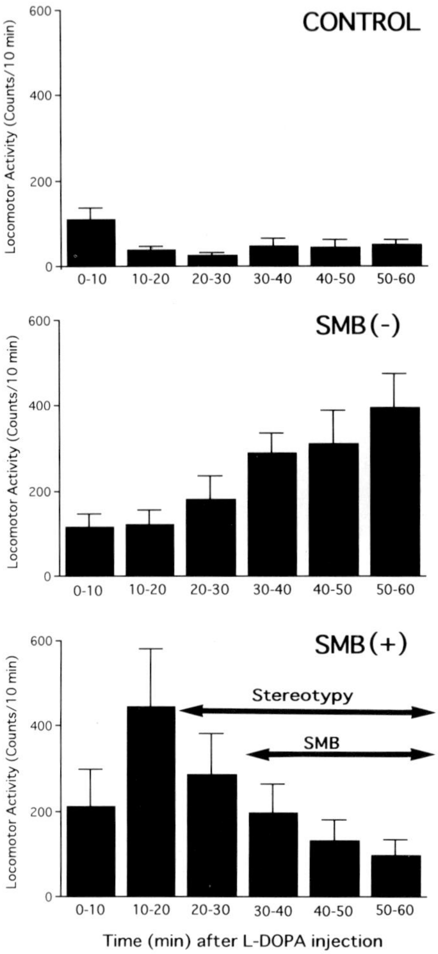

Fig. 2. Time dependent change of locomotor activity of neonatal 6-OHDA lesioned rats measured by Animex following L-DOPA loading in CONTROL, SMB(−), and SMB(+). In all instances, locomotor activity of the SMB(−) increased time dependently during 60 min. In SMB(+), locomotor activity suddenly increased, and thereafter stereotypy such as head nodding appeared, decreased locomotion for stereotypy, and finally SMB occurred. Mean ± SEM ($n = 10$).

loading, neonatal saline-treated control (CONTROL) rats showed slightly decreased locomotion counts. In contrast, the neonatal 6-OHDA treated group increased its locomotion regardless of SMB(+) and (−). However, the time schedules of the activity were significantly different from SMB(+) and SMB(−). SMB(−) rats steadily increased locomotor activity during 60 min following L-DOPA injection (delayed DOPA-induced increased locomotor activity: delayed DILA). However, they did not show SMB even if they showed 4–10 times higher locomotor counts than the untreated condition. SMB(+) rats rapidly increased their locomotor counts at a very fast phase following L-DOPA loading (fast DOPA-induced increased locomotor activity: fast DILA) and decreased their locomotor counts with the appearance of jumping or peculiar stereotypic behavior, head-nodding and bringing the mouth near their anterior extremities, then at last, SMB appeared. The fast DILA, peculiar stereotype, and SMB inevitably coexisted in SMB(+) rats. In some cases of SMB which occurred within 30 min, stereotypy and SMB immediately occurred without a prominent DILA. The association of fast DILA, stereotypy and SMB has not yet been determined.

The body weight gain was poor in neonatal 6-OHDA treated rats (*37*). Among the 6-OHDA treated group, SMB(+) rats tended to be smaller than SMB(−) rats.

These rats showed locomotor activity higher than normal rats when placed in a new environment. The total locomotor activity in the night phase increased in 6-OHDA-lesioned rats, but the locomotor counts among SMB(+) and SMB(−) did not change. There may be an impairment of the memory process since shuttle-box avoidance acquisition was impaired in neonatal 6-OHDA treated rats (*37*).

IV. MECHANISM OF SELF-MUTILATION BEHAVIOR IN NEONATAL 6-OHDA-LESIONED RAT: DOPAMINE D_1-LIKE RECEPTOR AND OTHER CANDIDATES

Pharmacobehavioral studies using specific agonists and antagonists have greatly contributed to our understanding of the role of dopamine in the manifestations of SMB. SMB in neonatally 6-OHDA-lesioned rats was induced by systemic loading of SKF-38393, a D_1

agonist, but systemic loading of quinpirole, a D_2 agonist, failed to induce it (*4*). Breese and his colleagues also found that L-DOPA induced SMB was inhibited by the systemic injection of SCH-23390, a D_1 antagonist, but was difficult to prevent by the systemic injection of haloperidol, a mixed antagonist of D_1- and D_2-like receptors (*4, 5*). From these findings it is evident that L-DOPA induced SMB through the D_1-like receptors. We also followed these studies, and further found that the D_2 specific antagonist, YM-09151-2, could not inhibit SMB.

Although the above behavioral studies suggest the supersensitivity of D_1-like receptors, the previous neurochemical studies failed to support this speculation. No studies found any elevation of D_1-like receptor bindings in the striatum in severely dopamine-depleted rats (*8, 16*). Dewar *et al.* (*15*) detected a slight increase of D_2-like receptor binding of striatal homogenates, but no change was observed in D_1-like receptor binding. The possibility of acceleration of D_1-mediated intracellular mechanisms was also tested. The D_1-like receptor stimulated adenylate cyclase activity was measured in striatal tissue, but no elevation of its activity has been found (*35*), although Luthman *et al.* (*25*) found the elevation of basal levels of striatal cyclic AMP. To date, neurochemical studies cannot account for the pharmacobehavioral discoveries of SMB in neonatal 6-OHDA lesioned rats.

Neonatal dopaminergic depletion influences the neurotransmission of other neurotransmitter systems in the rat brain. Lloyd *et al.* (*23*) speculated that the imbalance of transmitters of the striatum may cause a neurological symptom such as choleoathethosis or SMB in Lesch-Nyhan syndrome, because they found an increase of serotonin, a decrease of choline acetyltransferase activity and no change of GABA in addition to a severe depletion of dopamine. The serotonergic system is a potent candidate for the induction of SMB. This speculation is supported by the increase of serotonin content in the rat striatum (*5*), and hyperinnervation from dorsal raphe serotonergic neurons (*24, 38*) in neonatal 6-OHDA lesioned rats. Together with the finding that serotonergic receptors were upregulated in these rats (*31*), serotonergic neurotransmission was also increased. Interestingly, the increase of serotonin content was reported in the postmortem brain of Lesch-Nyhan disease (*23*), and the 5-hydroxytryptophan can attenuate the SMB in this disease (*27*). These findings suggest that SMB is related to serotonin neurotransmission, but we found no alteration of the serotonin content of caudate-putamen between SMB(+) and SMB(−). The effect of serotonergic agonistic and antagonistic drugs on the SMB should be examined in more detail in future.

Neonatal 6-OHDA treatment changes peptidergic neurotransmission in the striatum, in which many kinds of peptidergic neurons were distributed. Sivam (*36*) reported that the neonatal 6-OHDA increased the level of enkephalin but decreased the level of substance P. He speculated that substance P is important for the induction of SMB, since (1) substance P depletion was more severe following L-DOPA loading, (2) the change is inhibited by D_1 antagonist, and (3) substance P decrease was not observed in adult 6-OHDA lesioned rats. However, in his report, correlation of substance P levels and the behavioral SMB was not examined, thus it is still unclear whether or not the change is related to SMB. To learn the role of these peptides on SMB, behavioropharmacological studies using specific agonists and antagonists are needed.

V. MULTIPLE DOPAMINE NEURON SYSTEM AND SELF-MUTILATION BEHAVIOR

Since SMB is observed not temporarily but permanently in the rat model, perhaps permanent alteration of brain neuronal circuits may occur in these rats. The D_1-like receptor may be intimately linked to this neuronal circuit and thus be crucial for the induction of SMB. In the brain there are four major dopaminergic systems (nigro-dorsal striatal system, ventral tegmental area (VTA)-ventral striatal system, meso-cortical system, and intrahypothalamic system (*3*, *22*)), and through D_1-like and D_2-like receptors in these terminal fields, normal brain dopaminergic neurotransmission is conducted. Neonatal intracisternal or intraventricular 6-OHDA-treatment abolishes over 95% of the dopamine content in the brain (*5*, *37*). Then, are all dopaminergic neuron systems equally injured by neonatal 6-OHDA injection ? Which region of the dopamine receptor is altered following the destruction of dopaminergic neurons? If each dopaminergic neuron system shows different vulnerability to neonatal 6-OHDA toxicity, examination of the region of resistant and easily damaged

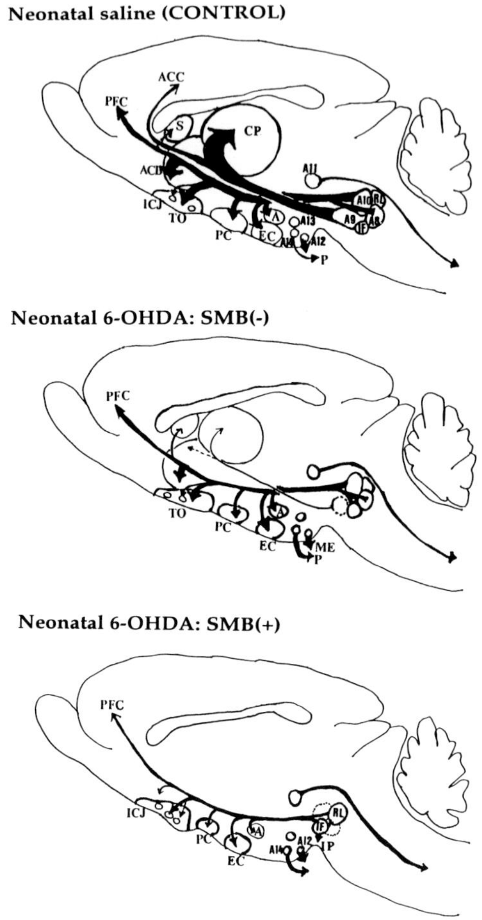

Fig. 3. Schematic representation of the dopaminergic neuron system detected by tyrosine hydroxylase immunocytochemistry in CONTROL, SMB(−), and SMB(+). Abbreviations: A, amygdaloid nuclei; ACB, nucleus accumbens; ACC, anterior cingulate cortex; CP, caudate-putamen; EC, entorhinal cortex; ICJ, insula of Calleja; IF, interfascicular nucleus; IP, interpeduncular nucleus; P, pituitary; PC, piriform cortex; PFC, prefrontal cortex; RL, rostral linear nucleus of raphe; S, septum lateralis; TO, tuberculum olfactorium; A8-A14, dopaminergic neuronal groups defined by Dahlstöm and Fuxe (*13*).

dopaminergic neurons may be of use in revealing the underlying dopaminergic system of SMB mediated by D_1-like receptors. By various morphological techniques, we tried to analyze the neuronal circuits involved in this abnormal behavior by comparing the neonatal 6-OHDA lesioned SMB(+) group, neonatal 6-OHDA lesions SMB(−) group, and control neonatal saline-treated group (CONTROL).

To find the dopamine-deficient region specific to SMB(+) rats, we employed immunocytochemistry using anti-tyrosine hydroxylase serum (Fig. 3). In control rats, dopamine neuron systems consist of (1) nigro-dorsal striatal system (substantia nigra-caudate putamen-cingulate cortex), (2) VTA-ventral striatal system (VTA-nucleus accumbens-tuberculum olfactorium), (3) meso-cortical system (midline midbrain nuclei-medial VTA-prefrontal cortex-amygdala-entorhinal cortex), and (4) intrahypothalamic system (arcuate-median eminence (A12), anterior hypothalamic periventricular area (A14), zone incerta (A13), and posterior hypothalamic area (A11). After the 6-OHDA treatment, the nigro-dorsal striatal system is easily destroyed in both SMB(+) and (−). The VTA-ventral striatal system is next susceptible to 6-OHDA toxicity, and the third, which is highly

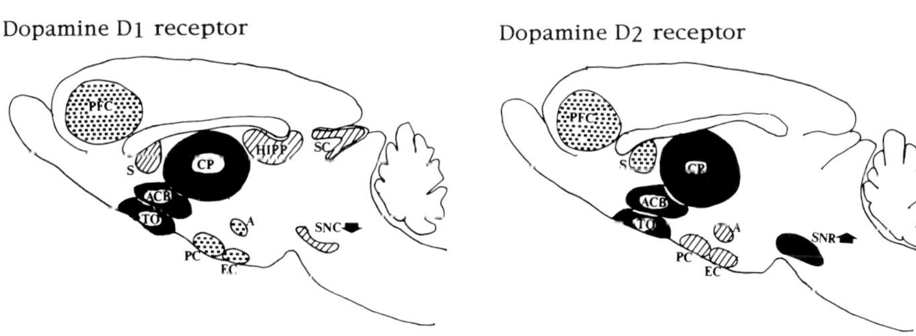

Fig. 4. Schematic representation of the dopamine D_1 and D_2 receptors in the rat brain by autoradiography. Binding densities were very high (>100 fmol/mg tissue) (●), high (100-40) (⊘), and moderate (40-20) (⊛) in D_1 by ^{3}H-SCH-23390, and very high (>60 fmol/mg tissue) (●), high (60-25) (⊘), and moderate (25-10) (⊛) in D_2 by ^{3}H-YM-09151-2. Although most D_1 and D_2 receptors were not altered in SMB(−) and SMB(+) compared to control, the increase of D_1 receptors in the substantia nigra pars reticulata (SNR) and the decrease of D_2 receptors in the substantia nigra pars compacta (SNC) were detected in SMB(+). For other abbreviations see the legend of Fig. 3.

resistant to death are the meso-cortical system and the intrahypothalamic system (*41, 42, 44*). This study demonstrates that the vulnerability of dopaminergic neurons to 6-OHDA toxicity is system-specific, and also shows the importance of the VTA-ventral striatal system for the SMB.

Then, does this region-specific damage accompany upregulation of region-specific dopaminergic receptors? To confirm this, we performed quantitative *in vitro* receptor autoradiography and receptor assays of homogenates using ^{3}H-SCH-23390 as D_1 ligand, and ^{3}H-YM-09151-2 as D_2 ligand (*43, 45*) (Fig. 4).

D_2-like receptor bindings of caudate-putamen tended to increase in the 6-OHDA-lesioned groups (SMB(−) and SMB(+)), but the differences among the three groups of CONTROL, SMB(−), and SMB(+) were not significant (Fig. 4; left panel). There was also no alteration of D_2-like receptor binding in the ventral striatum. D_2-like receptor binding in the pars compacta of the substantia nigra showing D_2 autoreceptor decreased in proportion to the decrease of dopaminergic cell number in SMB(−) and SMB(+). Quantitative *in situ* hybridization assay was performed using D_{2S} and D_{2L} specific oligonucleotide probes in the caudate-putamen, accumbens, and tuberculum olfactorium, but we found no statistically significant difference.

The D_1-like receptor antagonistic ligand, ^{3}H-SCH-23390 binds to striatal regions, limbic cortical area, and the substantia nigra pars reticulata (Fig. 4; right panel). There were no differences in the dorsal or ventral striatum among the CONTROL, SMB(−), and SMB(+). There was also no upregulation of D_1-like receptor, in spite of severe depletion of dopamine in SMB(+).

In the substantia nigra, the D_1-like receptor is known to be located in the terminals of the strio-nigral GABAergic tract (*2*). Although there was no difference in the binding between the CONTROL and SMB(−), SMB(+) showed an increase in binding. D_1 binding assay using the membrane preparation revealed that the Kd did not change, but Bmax of SMB(+) was significantly higher than SMB(−) and CONTROL (*29*). This indicates that the number of D_1-like receptors increased in the substantia nigra of SMB rat.

VI. A HYPOTHETICAL NEURONAL CIRCUIT

D_1-like receptor in the substantia nigra pars reticulata is known to receive dopamine released from dendrites of midbrain dopaminergic neurons (so-called dendritic release (*12*)), and to function as a regulator of GABA release from strio-nigral terminals (*39*). Since GABA is known to be a main transmitter regulating extrapyramidal output neurons in the pars reticulata, the upregulated D_1-like receptor of the substantia nigra may be an important functional point of L-DOPA and D_1 agonists for inducing the SMB phenomenon.

Intracerebral microinjection studies were performed to directly test the possible importance of the basal ganglia on the induction of SMB (Tables I and II). We implanted the cannula into bilateral striatum and bilateral substantia nigra of the rat brain which were previously confirmed to have SMB. After several days, we carried out

TABLE I
Inhibition Incidence of L-DOPA (100 mg/kg, i.p.)-induced SMB by Intracerebral Microinjection of Dopaminergic Antagonists

Injection site	SCH-23390 (5 μg/5 μl)	YM-09151-2 (10 μg/5 μl)
	Incidence of inhibition of SMB (positive/total)	
Caudate-putamen (hemi-injection)	0/6	0/6
Caudate-putamen (bilateral injection)	0/5	0/5
Substantia nigra (hemi-injection)	0/8	0/5
Substantia nigra (bilateral injection)	7/7	0/5

TABLE II
Occurrence Incidence of SMB by Intracerebral Microinjection of Dopaminergic Agonists

Injection site	SKF-38393 (1 μg/5 μl)	LY-141865 (1 μg/5 μl)	SKF 38393+LY-141865 (1 μg/5 μl+1 μg/5 μl)
		Incidence of SMB (positive/total)	
Caudate-putamen (hemi-injection)	0/3	0/4	0/3
Caudate-putamen (bilateral injection)	1/5	0/3	0/5
Substantia nigra (hemi-injection)	0/4	0/4	0/6
Substantia nigra (bilateral injection)	0/7	0/5	0/8
Bilateral caudate-putamen and bilateral substantia nigra	0/5	0/5	4/5

intraperitoneal L-DOPA loading, and SMB was induced. We microinjected D_1 antagonist SCH-23390 or D_2 antagonist YM-09151-2

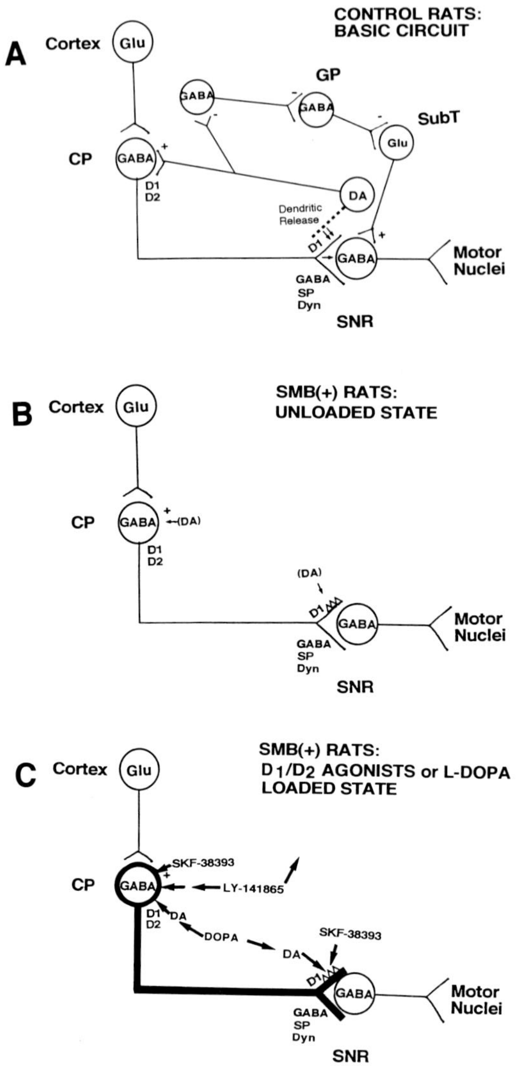

Fig. 5. Hypothetical figure explaining the induction and cessation of SMB by microinjection study. Basic neuronal circuits of basal ganglia are shown in uppermost panel (A), and in lower panels (B, C), strionigral GABAergic tract is emphasized. See text for further details.

unilaterally or both striatum or substantia nigra (Table I). Although we tried several injection combinations, only bilateral nigral SCH-23390 injection was successful. This indicates the striatal D_1-like receptor is not important for the cessation of SMB, but the upregulated nigral D_1-like receptor *is* crucial for the *cessation* of SMB.

Then, we searched for an important area for the *induction* of SMB. We injected dopamine receptor agonists to the striatum and the substantia nigra (Table II). We used SKF-38393 for D_1 agonist and LY-141865 for D_2 agonist. This clearly indicates that for the induction of SMB, the stimulation of D_1- and D_2-like receptors both in the substantia nigra and the striatum is needed. The stimulation of only the D_1-like receptor for all nigra and striatum is not enough for induction. This finding parallels the pharmacobehavioral studies of intraperitoneal injection of D_2 agonists and potentiates the induction of SMB induced by D_1 agonists (*7*). Moreover, it suggests the importance of striatum, in addition to the substantia nigra, in the *induction* of SMB.

Figure 5 is a hypothetical figure to explain the induction and cessation of SMB. Basic neuronal circuits of basal ganglia are shown in the uppermost panel, and in lower panels strionigral GABAergic tract is highlighted as an important factor for the induction and cessation of SMB. In normal rats, dopamine input increases the activity of strionigral GABAergic neurons (*1*). Dopaminergic dendritic release stimulates the release of GABA from strionigral terminals (*32*). GABA release from the terminals of the strionigral fibers produces inhibition of the tonically active neurons in the substantia nigra pars reticulata (*1*) that project to thalamus, superior colliculus, and pedunuclopontine nucleus.

In neonatal 6-OHDA rats non-loaded by dopamine agonists, dopamine neurons are severely injured, thus striatal stimulation and dendritic release of dopamine may decrease (Fig. 5B). However, the upregulated D_1-like receptor in the substantia nigra may compensate the neurotransmitter (GABA) release from nigral terminals. In rats showing SMB loaded by dopamine agonists or L-DOPA, strionigral neurons are activated by direct or indirect D_1 and D_2 agonists (Fig. 5C). Though the D_1 agonists directly bind to upregulated D_1-like receptor in the substantia nigra, the GABA release may be more potentiated and inhibit the extrapyramidal output neurons in the pars reticulata of the substantia nigra. Although the release of GABA has not been measured before in SMB state, Breese and his collaborators (*9*) reported that the injection of muscimol to substantia nigra induces SMB.

SUMMARY

Neurochemical and molecular approaches to complex behaviors such as this self-mutilation behavior are becoming popular. The present dopamine-deficiency model of SMB may be useful for resolving the mechanism, at least in some part. In this rat model, the only neurochemical factor that triggers SMB is dopamine. We analyzed the brain dopaminergic system focusing on regional differences. We demonstrated the destruction of both nigro-dorsal striatal system and VTA-ventral striatal system, and upregulation of nigral D_1 receptors in rats showing SMB. From these anatomical data, we made a brain microinjection study and found that the nigral D_1-like receptor is crucial for the cessation, and that both striatal and nigral D_1- and D_2-like receptors are important for the induction. However, no data are available on the factor determining the characteristics of the behavior. As one possible candidate, the association of the impairment of limbic striatal function, which is suggested by the destruction of the VTA-ventral striatal dopaminergic neuron system, and SMB is interesting. Although SMB has various clinical expressions and is not caused by a simple neurological background, a trial to elucidate the neuronal circuits of this experimental model may help find a therapy for this condition.

REFERENCES

1 Alexander, D.E. and Crutcher, M.D. (1990). *Trends Neurosci.* **13**, 266–271.
2 Altar, C.A. and Hauser, K. (1987). *Brain Res.* **410**, 1–11.
3 Bjöklund, A. and Lindvall, O. (1984). In *Handbook of Chemical Neuroanatomy*, ed. Bjöklund, A. and Hökfelt, T., vol. 2, pp. 55–122. Amsterdam: Elsevier.
4 Breese, G.R., Baumeister, A.A., Napier, T.C., Frye, G.D., and Mueller, R.A. (1985). *J. Pharmacol. Exp. Ther.* **235**, 287–295.
5 Breese, G.R., Baumeister, A.A., McCown, T.J., Emerick, S.G., Frye, G.D.,

Crotty, K., and Mueller, R.A. (1984). *J. Pharmacol. Exp. Ther.* **235**, 287–295.
6 Breese, G.R., Baumeister, A.A., McCown, T.J., Emerick, S.G., Frye, G.D., and Mueller, R.A. (1984). *Pharmacol. Biochem. Behav.* **21**, 459–461.
7 Breese, G.R., Criswell, H.E., Duncan, G.E., and Mueller, R.A. (1990). *Brain Res. Bull.* **25**, 477–484.
8 Breese, G.R., Duncan, G.E., Napier, T.C., Bondy, S.C., Iorio, L.C., and Mueller, R.A. (1987). *J. Pharmacol. Exp. Ther.* **240**, 167–176.
9 Breese, G.R., Hulebak, K.L., Napier, T.C., Baumeister, A.A., Frye, G., and Mueller, R.A. (1987). *Psychopharmacology* **91**, 356–362.
10 Breese, G.R., McCown, T.J., Baumeister, A.A., Emerick, S., Frye, G.D., and Mueller, R.A. (1984). *Fed. Proc.* **43**, 928.
11 Bruno, J.P., Stiker, E.M., and Zigmond, M.J. (1985). *Behav. Neurosci.* **99**, 771–775.
12 Cheramy, A., Leviel, V., and Glowinski, J. (1981). *Nature* **289**, 537–542.
13 Dahlstöm, A. and Fuxe, K. (1964). *Acta Physiol. Scand.* **62**, (Suppl.) **232**, 1–55.
14 Delbarre, F., Cartier, P., Auscher, C., De Géry, A., and Hamer, M. (1970). *Pres. Méd.* **78**, 729–734.
15 Dewar, K.M., Soghomonian, J.-J., Bruno, J.P., Descarries, L., and Reader, T.A. (1990). *Brain Res.* **536**, 287–296.
16 Duncan, G.E., Greese, G.R., Criswell, H.E., Johnson, K.B., Schambra, U.B., Mueller, R.A., Caron, M.G., and Fremeau, R.T.J. (1993). *J. Neurochem.* **61**, 1255–1262.
17 Fink, J.S., Weaver, D.R., Rivkees, S.A., *et al.* (1992). *Mol. Brain Res.* **14**, 186–195.
18 Goldstein, M., Kuga, S., Kusano, N., Meller, E., Dancis, J., and Schwarcz, R. (1986). *Brain Res.* **367**, 114–120.
19 Jinnah, H.A., Gage, F.H., and Friedmann, T. (1990). *Brain Res. Bull.* **25**, 467–475.
20 Jinnah, H.A., Wojcik, B.E., Hunt, M., Narang, N., Lee, K.Y., Goldstein, M., Wamsley, J.K., Langlais, P.J., and Friedmann, T. (1993). *J. Neurosci.* **14**, 1164–1175.
21 Lesch, M. and Nyhan, W.L. (1964). *Am. J. Med.* **36**, 561–570.
22 Lindvall, O. and Björklund, A. (1974). *Acta Physiol. Scand.* **412**, (Suppl.), 1–48.
23 Lloyd, K.G., Hornykiewicz, O., Davidson, L., Shannak, K., Farley, I., Goldstein, M., Shibuya, M., Kelley, W.N., and Fox, I.H. (1981). *N. Engl. J. Med.* **305**, 1106–1111.
24 Luthman, J., Bilioli, B., Tsutsumi, T., Verhofstad, A., and Jonsson, G. (1987). *Brain Res. Bull.* **19**, 269–274.
25 Luthman, J., Lindqvist, E., Young, D., and Cowburn, R. (1990). *Exp. Res. Bull.* **19**, 85–95.
26 Marks, J.F., Baum, J., Kay, J.L., Taylor, W., and Curry, L. (1968). *Pediatrics* **42**, 357–361.
27 Mizuno, T. and Yurari, Y. (1975). *Neuropädiatrie* **6**, 13–23.
28 Nyhan, W.L. (1973). *Annu. Rev. Med.* **24**, 41–60.

29 Okamura, H., Yokoyama, C., Kawakami, F., and Ibata, Y. (1993). *Soc. Neurosci. Abstr.* **19**, 1245.

30 Page, T., Bakay, B., Nissinen, E., and Nyhan, W.L. (1981). *J. Inher. Metabol. Dis.* **4**, 203–206.

31 Radja, F., Descarries, L., Dewar, K., and Reader, T.A. (1993). *Brain Res.* **606**, 273–285.

32 Reubi, J.-C., Iversen, L.L., and Jessell, T.M. (1977). *Nature* **268**, 652–654.

33 Sandy, K.R. (1987). *J. Child Neurol.* **3**, 147–148.

34 Shaywitz, B.A., Yager, R.D., and Klopper, J.H. (1976). *Science* **191**, 305–308.

35 Simson, P.E., Johnson, K.B., Jurevics, H.A., Criswell, H.E., Napier, T.C., Duncan, G.E., Mueller, R.A., and Breese, G.R. (1992). *J. Pharmacol. Exp. Ther.* **263**, 1454–1463.

36 Sivam, S.P. (1989). *Brain Res.* **500**, 119–130.

37 Smith, R.D., Cooper, B.R., and Breese, G.R. (1973). *J. Pharmacol. Exp. Ther.* **185**, 609–619.

38 Snyder, A.M., Zigmond, M.J., and Lund, R.D. (1986). *J. Comp. Neurol.* **245**, 274–281.

39 Waszczak, B.L. and Walters, J. (1986). *J. Neurosci.* **6**, 120–126.

40 Winchel, R.M. and Stanley, M. (1991). *Am. J. Psychiatr.* **148**, 306–317.

41 Yagita, K., Okamura, H., Yokoyama, C., Taguchi, J., Nagatsu, I., and Ibata, Y. (1991). *Acta Histochem. Cytochem.* **24**, 538.

42 Yokoyama, C., Okamura, H., and Ibata, Y. (1993). *Brain Res. Bull.* **30**, 551–559.

43 Yokoyama, C., Okamura, H., and Ibata, Y. (1995). *Brain Res.* **681**, 153–159.

44 Yokoyama, C., Okamura, H., Nakajima, T., and Ibata, Y. (1993). *Soc. Neurosci. Abstr.* **19**, 1245.

45 Yokoyama, C., Okamura, H., Nakajima, T., Taguchi, J., and Ibata, Y. (1994). *J. Comp. Neurol.* **344**, 121–136.

17
PROSTAGLANDINS AND SLEEP

OSAMU HAYAISHI

Osaka Bioscience Institute, Suita, Osaka 565, Japan

Although we repeat sleep-wake cycles every day and night, and spend almost one-third of our lifetime sleeping in bed or somewhere else, the molecular mechanisms of sleep-wake regulation have thus far been hardly understood. Sleep is one of the most important and yet least understood biological phenomena of the brain. In spite of the intensive effort by numerous investigators since the beginning of this century, we still cannot answer simple questions such as: what is sleep?, why do we need to sleep?, how is sleep regulated?, and so forth. In this review, I shall attempt to summarize our present and past work in three short and seemingly simple stories; first, prostaglandin (PG) D_2 makes you sleep. Second, PGD synthase is the key enzyme in sleep regulation and third, prostaglandin E_2 makes you wake.

The concept of humoral rather than neural regulation of sleep was initially proposed by a French neuroscientist, Henri Pieron of Paris in the first decade of this century. He and his associate Legendre were the first to show the presence of a sleep substance in the cerebrospinal fluid (CSF) of sleep-deprived dogs. Concurrently and independently, Kuniomi Ishimori of Nagoya University, Nagoya, Japan, employing a similar experimental approach, also demonstrated a sleep-inducing substance in the CSF of sleep-deprived dogs. However, the chemical nature of their sleep substances was not identified. During the next 80 years or so, more than 30 so-called endogenous sleep substances have been reported from brains, cerebrospinal fluid, and other organs and tissues of mammals by numerous investigators. However, their physiological relevance has remained uncertain in most instances.

I. PGD_2 INDUCES PHYSIOLOGICAL SLEEP

By way of introduction, I would first like to say a few words about sleep and how sleep is determined. Wakefulness and two major types of sleep, namely, slow-wave sleep or SWS and rapid eye movement or REM sleep, can be judged by the animal's behavior. However, more precise qualitative and quantitative assessment of the sleep-wake pattern is made by the EEG, or electroencephalogram, namely the recording of brain waves; the EOG, or electro-oculogram, the recording of eye movement; and by the EMG, or electromyogram, which is a recording of muscle tension. The bioassay system currently in use in my laboratory is a modified version of the continuous infusion sleep bioassay system originally developed by Professor S. Inoué and his associates. Samples are infused continuously and slowly through a chronically implanted cannula by means of a microinjection pump into the third ventricle of a freely moving rat. The sleep stages are determined on the basis of polygraphic recordings of EEG and EMG. Brain temperature as well as food and water intake is also monitored, and behavior of the rat is recorded by a video-recording system.

Rats are nocturnal animals and normally sleep during the day, but are awake most of the time during the night. When PGD_2 was infused at a rate of 0.6 pmol per min during the night, both SWS and REM sleep increased significantly. The effect was specific for PGD_2 and dose-dependent. As little as one femtomole (10^{-15} mol) of PGD_2 per sec was effective in inducing excess sleep. Most importantly, sleep induced by PGD_2 was indistinguishable from physiological sleep as judged by EEG, EMG, locomotor activities, brain temperature, heart rate, and general behavior of the rat.

In order to further prove that the effect of PGD_2 was not pharmacological, but physiological, we determined the specific activity of the enzyme that produces PGD_2 in the rat brain, namely PGD synthase, during the sleep-wake cycle. The PGD synthase activity in the rat brain exhibited a circadian fluctuation in parallel with the sleep/wake cycle. In contrast, the specific activities of other related enzymes, for example, PGD_2 dehydrogenase, remained almost the same during the day and night. More recently, Pandey and Ram in my laboratory showed that the concentration of PGD_2 itself in the rat CSF also exhibited a similar circadian fluctuation in parallel with the sleep/wake cycle (unpublished observation). If PGD synthase is indeed involved in the regulation of physiological sleep, when the synthase activity during the day is inhibited by a specific enzyme inhibitor, sleep should also be inhibited and the amount of diurnal sleep should decrease. In order to find a specific inhibitor of PGD synthase, we then purified this enzyme and studied its properties.

II. PGD SYNTHASE

PGD synthase catalyzes the isomerization of the substrate PGH_2 to produce PGD_2. This enzyme is a monomeric glycoprotein with a molecular weight of approximately 27,000. Urade and co-workers in my laboratory purified the enzyme from human and rat brains, isolated cDNAs encoding PGD synthase, and determined the nucleotide and amino acid sequences (*8*). These experiments provided several lines of important and unexpected information, which opened up a new insight into our understanding of the mechanism of the somnogenic activity of PGD_2.

By a computer-assisted homology search in data bases of protein primary structure, PGD synthase was shown to be a member of the lipocalin superfamily. Members of this superfamily are widely distributed in the animal kingdom, being found in a variety of mammals, birds, amphibians, insects, and crustacea. These proteins are small secretory proteins sharing a common feature for the binding and transport of small lipophilic molecules. The only exception to this rule is PGD synthase, which is an enzyme rather than a lipid transporter. The gene structure of the members of this family has also been delineated, and that of the rat enzyme was found to be remarkably analogous to the gene structures of several other members of this family, such as α-2-urinary globulin, β-lactoglobulin, placental protein 14, *etc.*, in terms of number and sizes of exons and phase of splicing of introns. Based on these results, a phylogenic tree of the lipocalin superfamily was constructed by Dr. H. Toh of the Protein Engineering Institute. Because of the high evolutionary divergence of this superfamily, the homology of the amino acid sequences of the members is rather weak; yet the tertiary structure is well conserved to form a remarkably similar β-barrel structure, as revealed by X-ray crystallographic studies of retinal-binding protein, β-lactoglobulin, and other members of this family.

The tertiary structure of the brain PGD synthase was computer-graphically deduced. The free SH group due to cysteine residue 65 is found only in the human and rat PGD synthase and not in the other lipocalins. It is in the hydrophobic pocket and is considered to be in the active site of the enzyme. In fact, replacement of cysteine 65 by serine or alanine by site directed mutagenesis resulted in complete loss of the enzyme activity. Islam and co-workers in my laboratory (*4*) discovered that inorganic quadrivalent (4^+) selenium compounds are potent, specific, noncompetitive, and reversible inhibitors of brain PGD synthase, presumably interacting with this free sulfhydryl group in the active center. *In vitro* kinetic studies revealed that various quadrivalent selenium compounds are inhibitory, with K_i values of about 10 μM. However, hexavalent and organic selenium compounds are hardly effective. This inhibition is reversible and can be reversed by the addition of excess amounts of sulfhydryl compounds such as glutathione or 1,4-dithiothreitol (DTT). Other enzymes in the arachidonate cascade system are not inhibited by any of these selenium compounds. We then administered these selenium compounds into the third ventricle of a freely moving rat to see if they would have any inhibitory effect on sleep. When selenium chloride was infused into the third ventricle of a rat during the day, sleep was inhibited time- and dose-dependently. About 2 hr after the start of the infusion, both SWS and REM sleep were almost completely inhibited. The effect was reversible. When the infusion was interrupted, sleep was restored. Furthermore, the effect was reversed by the simultaneous infusion of sulfhydryl compounds such as DTT and reduced glutathione, as in the case of the *in vitro* enzyme activity. These results therefore clearly show that the PGD synthase is the key enzyme in sleep regulation and that PGD_2 is the endogenous sleep-regulating substance in the rat under physiological conditions (the earlier results are reviewed in (*2*) and (*3*)).

III. LOCALIZATION OF PGD SYNTHASE

If this be the case, questions arise like: "in what region is the PGD synthase located in the mammalian brain?" and "in what type of cells?" In an attempt to answer these questions, we employed three independent approaches, namely, *in situ* hybridization, immunohistochemical staining and the enzyme assay. The results were quite remarkable and unexpected. Typical results of *in situ* hybridization of a rat brain and spinal cord clearly indicate that mRNA for PGD synthase is expressed faintly and diffusely in the white matter of the brain. However, much stronger signal is seen in the leptomeninges, namely pia mater and arachnoid membrane, and choroid plexus in the ventricles. Leptomeninges and choroid plexus are loaded with PGD synthase mRNA and enzyme protein. Oligodendrocytes are also positively stained, but little activity is seen in other cells including neurons (*12*).

Based on these morphological observations, we then determined the enzyme activity in these tissues and also in the CSF of rats and humans. The results were consistent with those obtained with *in situ* hybridization and immunohistochemistry. The specific activities in the leptomeninges and choroid plexus were much higher than that in the whole brain. Furthermore, both rat and human CSF contains a remarkably large amount of PGD synthase activity. Recently, Dr. C. Murphy of the University of Tennessee, Knoxville isolated a protein from human urine, the partial sequence of which was identical to that of PGD synthase. This urinary protein was also shown to have a significant enzyme activity. β-Trace, which is the major protein in CSF showed a very high specific activity, almost comparable to the most purified enzyme preparation of rat brain. β-Trace is the second most abundant protein in human CSF after albumin. It was discovered in 1961 as a protein specific to CSF, but its function has not yet been elucidated. Recently several groups of investigators in Germany reported that the N-terminal partial sequence of human β-trace is highly homologous to that of human brain PGD synthase except that β-trace has no signal peptide. We then purified β-trace from human CSF and found that β-trace and PGD synthase were structurally, enzymatically, and immunologically identical (*13*).

IV. SITE OF ACTION OF PGD_2

In an attempt to determine the exact site of action of the sleep-promoting effect of PGD_2, Matsumura and co-workers recently infused a small amount of PGD_2 in nearly 300 different sites in the rat brain. To our great surprise, bilateral infusion of PGD_2 into the subarachnoid space in a surface area of the medial ventral region of the rostral basal forebrain was found to be most effective. SWS was increased more than 120 min above the baseline value of 100 min during the 6 hr of PGD_2 infusion (*7*). It has been known for some time that the chemosensitive regions involved in the regulation of blood pressure and respiration are located in the ventrolateral surface region of the brain stem, and these regions have been termed rostral, intermediate, and caudal areas by Loeschcke and co-workers (*5*). These chemosensors are reportedly activated by various ligands, such as hydrogen ion, CO_2, acetylcholine, serotonin, endothelin, and other compounds. It would not be surprising, therefore, to find a chemosensitive locus for PGD_2 in these surface areas. All these experimental observations taken together led us to postulate the following tentative conclusion.

PGD synthase, the enzyme that produces PGD_2 in the rat brain, is present mainly, if not exclusively, in the membrane system surrounding the brain rather than the brain parenchyma, and is secreted into the CSF to become β-trace protein, a major protein component of the CSF. PGD synthase as well as the PGD_2 thus produced circulates in the ventricular system, subarachnoidal space, and extracellular space in the brain system. PGD_2 interacts with chemosensors or receptors on the ventro-medial surface of the rostral basal forebrain, thus initiating the signal to make the brain sleep.

Recently, two principal modes of electrochemical communication in the brain, "volume transmission through the brain extracellular fluid and CSF", as opposed to the conventional "wiring transmission through the nerve cells and synapses", were advocated by Fuxe and Agnati of the Karolinska Institute (*1*). The volume transmission was suggested to have advantages for global regulations such as consciousness, sleep, pain, emotion, and so forth. Our findings are consistent with their concept in the sense that PGD_2 is not a typical neurotransmitter but rather a "neurohormone" or an "informational substance" that circulates through the CSF in the ventricular system and the extracellular space and transmits certain chemical messages.

V. STUDIES WITH MONKEYS AND HUMANS

We further extended these studies to the *Macaca mulatta*, that is, the rhesus monkey. Essentially similar results were obtained indicating that PGD_2 induces physiological sleep in primates.

So far I have emphasized that the sleep induced by PGD_2 is indistinguishable from natural sleep and that PGD_2 controls physiological sleep in rats and monkeys. How, then, is pathological sleep controlled? Are human sleep disorders, such as sleeping sickness, narcolepsy and so forth caused by aberrant metabolism of prostaglandins? In 1990, Pentreath and co-workers at Sulford University, U.K., reported that PGD_2 concentration is markedly (10^2–10^3-fold), progressively and selectively elevated in the CSF of patients with advanced sleeping sickness induced by *Trypanosoma brucei gambiense*, while PGE_2 and interleukin-1 (IL-1) levels did not rise significantly (*10*). Roberts and Oates at Vanderbilt University, U.S.A., reported that profound lethargy, namely, deep sleep exhibited by systemic mastocytosis patients, is primarily due to the episodic release of mast cell mediators (*11*). Up to 150-fold increases in endogenous production of PGD_2 have been found in patients with mastocytosis. Based on these accumulating experimental evidence and clinical observations reported from my own and other laboratories, it is reasonable to conclude that PGD_2 induces sleep in humans also, and that certain pathological types of sleep are caused by excessive endogenous production of PGD_2. All of these results taken together clearly show that PGD_2 is the endogenous sleep-regulating substance under physiological as well as pathological conditions and that PGD synthase is the key enzyme in sleep regulation.

VI. WAKING EFFECT OF PGE_2

Now, I would like to briefly describe the waking effect of PGE_2. PGE_2 and PGD_2 are positional isomers and have been shown to sometimes exhibit opposite biological activities. For example, PGD_2 lowers body temperature, suppresses secretion of luteinizing hormone-releasing hormone, and decreases ion transport across rat colon mucosal membrane, while PGE_2 shows the opposite effects (*6*).

In examining the neural functions of PGE_2, we found that intraventricular infusion of PGE_2 reduced the amount of sleep or increased the time of wakefulness. When PGE_2 was infused at a rate of 10 pmoles/min into the third ventricle of a rat during the day, the amounts of both SWS and REM sleep were reduced significantly. After the termination of infusion, there was a rebound, and nocturnal sleep was increased to a significant extent but the sleep pattern returned to normal during the next day, indicating that the effect of PGE_2 is not due to its toxicity.

In order to find out if PGE_2 is indeed involved in the regulation of the sleep-wake cycle, we examined the effect of AH6809, a Glaxo product, that has been established as a PGE_2 antagonist in smooth muscle relaxation and the pain system. If PGE_2 induces wakefulness or inhibits sleep under physiological conditions, such a PGE_2 antagonist should counteract the effect of endogenous PGE_2; that is, it should increase the amount of sleep or decrease the time of wakefulness. When the PGE_2 antagonist was infused at a rate of 20 pmoles/min into the third ventricle of a rat during the night, the amounts of both SWS and REM sleep increased. Such an experiment clearly indicates that PGE_2 is involved in the maintenance of the waking state under physiological conditions.

To find the site of action of PGE_2, Onoe in my laboratory infused various amounts of PGE_2 into different areas of the monkey brain using a micro-dialysis probe and then determined its effects on brain temperature and wakefulness. The results indicated that the site of action of PGE_2 for brain temperature regulation is near the preoptic, anterior hypothalamic region and that that for wakefulness is near the tuberomammillary nucleus region in the posterior hypothalamus (*9*). It should be pointed out, however, that the amount of REM sleep was hardly affected under these conditions, but the infusion of PGE_2 into the aqueduct decreased the amount of REM sleep. These findings indicate that the putative center for REM sleep regulation in the brain stem is also under the regulatory control of PGE_2.

SUMMARY

PGD_2 and E_2 are the major endogenous sleep-regulating substances, the former inducing physiological sleep and the latter wakefulness in rats, monkeys and probably in humans as well. Second, PGD synthase, the key enzyme in the regulation of sleep, is present mainly in the leptomeninges and choroid plexus, from which it is secreted into the CSF to become β-trace protein. Third, PGD_2 exerts its somnogenic activity at the ventro-rostral surface of the basal forebrain, whereas the site of action of PGE_2 appears to be in or near the posterior hypothalamus.

Needless to say, sleep is a very complex phenomenon, and these prostaglandins are only two among a large number of gears involved in the control of sleep and wakefulness, but they do seem to play a major role in the initiation and maintenance of the sleep propensity. Studies are in progress in my own and other laboratories concerning the regulatory mechanisms of PGD_2 biosynthesis and the molecular mechanisms involved in the transmission of the message initiated by those prostaglandins. We realize that we have a long way to go to attain our goal of understanding in biochemical terms the entire mechanism of sleep-wake regulation. I hope we are on the right track.

REFERENCES

1 Fuxe, K. and Agnati, L.F. (eds.) (1991). *Volume Transmission in the Brain.* New York: Raven Press

2 Hayaishi, O. (1988). *J. Biol. Chem.* **263**, 14593–14596.

3 Hayaishi, O. (1991). *FASEB J.* **5**, 2575–2581.
4 Islam, F., Watanabe, Ya., Morii, H., and Hayaishi, O. (1991). *Arch. Biochem. Biophys.* **289**, 161–166.
5 Loeschcke, H.H. (1982). *J. Physiol.* (Lond.) **332**, 1–24.
6 Matsumura, H., Goh, Y., Ueno, R., Sakai, T., and Hayaishi, O. (1988). *Brain Res.* **444**, 254–272.
7 Matsumura, H., Nakajima, T., Osaka, T., Satoh, S., Kawase, K., Kubo, E., Sri Kantha, S., Kasahara, K., and Hayaishi, O. (1994). *Proc. Natl. Acad. Sci. U.S.A.* **91**, 11998–12002.
8 Nagata, A., Suzuki, Y., Igarashi, M., Eguchi, N., Toh, H., Urade, Y., and Hayaishi, O. (1991). *Proc. Natl. Acad. Sci. U.S.A.* **88**, 4020–4024.
9 Onoe, H., Watanabe, Ya., Ono, K., Koyama, Y., and Hayaishi, O. (1992). *J. Neurosci.* **12**, 2715–2725.
10 Pentreath, V.W., Rees, K., Owolabi, O.A., Philip, K.A., and Doua, F. (1990). *Trans. Royal Soc. Trop. Med. Hyg.* **84**, 795–799.
11 Roberts, J.L., II, Sweetman, B.J., Lewis, R.A., Austen, K.F., and Oats, J.A. (1980). *New Engl. J. Med.* **303**, 1400–1404.
12 Urade, Y., Tanaka, T., Eguchi, N., Kikuchi, M., Kimura, H., Toh, H., and Hayaishi, O. (1995). *J. Biol. Chem.* **270**, 1422–1428.
13 Watanabe, K., Urade, Y., Mäder, M., Murphy, C., and Hayaishi, O. (1994). *Biochem. Biophys. Res. Commun.* **203**, 1110–1116.

SUBJECT INDEX